Solomon Maimon

MONISM
SKEPTICISM
AND
MATHEMATICS

Solomon Maimon

Meir Buzaglo

UNIVERSITY OF PITTSBURGH PRESS

Published by the University of Pittsburgh Press, Pittsburgh, Pa., 15260

Manufactured in the United States of America

Printed on acid-free paper

10 9 8 7 6 5 4 3 2 1

Library of Congress Cataloging-in-Publication Data

Buzaglo, Meir, 1959–
Solomon Maimon : monism, skepticism, and mathematics / Meir Buzaglo.
p. cm.
Includes bibliographical references and index.
ISBN 0-8229-4176-7 (alk. paper)
1. Maimon, Salomon, 1754–1800. I. Title.
B3068 .B89 2002
181'.06—dc21

2001006537

TO MY BELOVED BROTHER SHALOM

Contents

Preface

The philosophy of Solomon Maimon is usually considered an important link between Kant's transcendental philosophy and German idealism. Here I attempt to read Maimon as an independent post-Kantian thinker. Maimon's importance does not derive from the greatness of the systems he undoubtedly influenced; his is a lasting contribution in its own right. An understanding of Maimon's work can significantly contribute to an understanding of certain complex notions in Kant's work, such as intuition, affection, and the thing-in-itself. Maimon's work also gives us a better understanding of the system of Fichte and Hegel; however, its main merit lies in suggesting a system that combines idealism and rationalism in an original way.

This study is essentially a reconstruction, from an analytical perspective, of the system Solomon Maimon proposed as a challenge to Kant's concept-intuition dualism. In proposing such a reconstruction, I am, in effect, carrying out the detailed plan outlined by Maimon himself in his *Versuch über Transcendentalphilosophie,* a draft of which was highly acclaimed by Kant. This plan is not developed in Maimon's later work.

Although Kant rejected the idea of a given without a conceptual element, his system is distinctly dualistic. The starting point for the dialogue between Kant and Maimon is the question of the possibility of a connection between intuition and concept. The originality of Maimon lies in his translation of this problem into the mathematical realm. He assumed that in mathematics sensibility is not independent of understanding, and intuition is nothing but the concept in its potential mode. This answer is the basis for introducing Maimon's synthesis among Hume, Kant, and Leibniz. I address the two central problems in reading Maimon. The first is making the connection between sensibility and understanding in the mathematical realm. The second is determining whether and how it can be extended to knowledge in general.

One difficult issue stems from the intimate connection between Maimon's philosophy and that of Kant. The Kantian scholar is liable to stray into hermeneutic queries on the text and questions such as whether Maimon understood Kant or whether Kant provides the answer to Maimon's misgivings, and, of course, questions about my own interpretation of Kant are bound to arise. But I have refrained from examining such issues in depth due to the need to provide a clear, concise picture of Maimon's philosophy.

Acknowledgments

I began writing on Maimon before I knew that the discussion between analytic philosophy and Kantian heritage was mainstream in philosophy. Truth be told, after graduating from the department of mathematics, I decided to work on Maimon *because* I thought that rationalism and idealism were not in fashion. Fortunately, however, it turns out that Maimon belongs to the heart of the multidimensional dialogue between Kant's philosophy and modern thought.

My interest in Maimon began with the discovery of his autobiography in my brother Shalom's library. I was a sixteen-year-old boy. I owe Shalom my interest in philosophy (and much more), and that is why this book is dedicated to him. I wish to thank Samuel Scolnicov, who as a student of Bergman encouraged me to work on Maimon and was my first mentor. Special thanks to my teacher and mentor Gilead Bar-Elli, who agreed to supervise my work on Maimon. I met Gilead at a crucial point of my studies, at a time when I had not yet chosen a style of doing philosophy. Without his reading and advice I am sure the unity of this book would be weak and unclear. Avital Wohlman read an early version of the manuscript and made helpful comments. She doesn't know how supportive the evenings were when we went over portions of this work. Eva Shorr was very helpful in providing answers to my endless questions on editing and preparing the manuscript for publication. I wish to thank the students of my seminars on Maimon at the Hebrew University and my assistant, Mr. Adam Sager, for helping me at different stages of my work.

I owe an apology to my daughter, Neria, for the time I was not there for her. I can only say she is the reader I have imagined. Finally, I wish to thank my wife, Dina, for many things, which I prefer to tell her privately.

Abbreviations

Versuch	Solomon Maimon. *Essay on Transcendental Philosophy* (Versuch über transcendentalphilosophie: Mit einem Anhang über die symbolische Erkenntnis und Anmerkungen). 1790. Vol. 2 of Maimon, *Gesammelte Werke* (Hildesheim: Georg Olms Verlagsbuchhandlung, 1965).
Logik	Solomon Maimon. *Attempt at a New Logic or a Theory of Thought* (Versuch einer neuen logik oder theorie des denkens: Nebst angehangten briefen des Philaltes an Aenesidemus). 1794. Vol. 5 of Maimon, *Gesammelte Werke* (Hildesheim: Georg Olms Verlagsbuchhandlung, 1965).
Autobiography	Solomon Maimon. *The Autobiography of Solomon Maimon.* (Lebensgeschichte). 1792. Trans. from the German by J. Clark Murray. London: East and West, 1954. Vol. 1 of Maimon, *Gesammelte Werke* (Hildesheim: Georg Olms Verlagsbuchhandlung, 1965).
Streiferein	Solomon Maimon. *Streiferein in Gebiete der Philosophie.* 1793. Vol. 4 of Maimon, *Gesammelte Werke* (Hildesheim: Georg Olms Verlagsbuchhandlung, 1965).
Critique	Immanuel Kant. *Critique of Pure Reason. A* refers to the first edition and *B* to the second edition (see bibliography).
Letters	Immanuel Kant. *Philosophical Correspondence, 1759–1799.* Ed. and trans. by A. Zweig. Chicago: University of Chicago Press, 1967.

Solomon Maimon

Introduction

THAT MAIMON'S life was not the happiest is clearly attested in his autobiography—a tale of wandering and misfortune. Yet tragedy in a different sense attaches to his main composition, *Versuch über Transcendentalphilosophie (Versuch)*. The book was not understood by his contemporaries. The editor of the *Algemeine Literature Zeitung* wrote Maimon that he had given it to three "of the most profound" scholars and they refused to pass an opinion on the work. Karl Leonard Reinhold, who had assumed the role of Kant's official interpreter and who eventually became Maimon's opponent, confessed that for two entire years he tried to understand Maimon but without success. The book was perceived as lacking a plan and brimming with contradictory notes.

Kant and Fichte were two exceptions in this regard. Kant praised Maimon, saying, "none of my critics have understood me as Maimon" (Kant to Marcus Herz, 26 May 1789, *Letters* 151). Fichte expressed "boundless respect for Maimon's talent" (Bergman 1967, 229), implicitly censuring his colleagues for having ignored Maimon. Nevertheless, even Kant, at a later stage, turned his back on Maimon and declared: "As regards the 'improvement' of the critical philosophy by Maimon . . . I have never really understood what he was after and must leave the reproof to others" (Kant to Reinhold, 28 March 1794, *Letters* 211). On second inspection, it transpires that even in the first letter, where Kant expressed his appreciation of Maimon, Kant did not really understand him. He seemed to have sensed that there was something to Maimon's contention, but he did not quite know what.

The mist shrouding the *Versuch*, Maimon's most important work, did not dissipate in subsequent generations, and in fact it has endured to this day. Atlas (1964, 10) writes that Maimon's is perhaps the obscurest of all philosophical writings. The interpretative questions, as Beiser (1987, 287) has made clear, concern what Maimon is actually trying to say. Where does his importance lie? Is it in his unique construal of thing-in-itself (Bergman 1968)? Is it in his theory of fiction (Vaihinger 1925)? Was it Maimon's aim to force the dilemma between rationalism and skepticism (Atlas 1964)? Did he wish to blaze a midway between Hume and Leibniz (Cassirer 1920; Beiser 1987)? How are we to understand his obscure theory of differentials? What about the infinite understanding? These are a few of the fundamental questions regarding which there is no consensus in Maimon's scholarship. Moreover, there is also no definite debate between the commentators, nor is there an agreed demarcation of what is central to the system, for it is not at all clear that there is a Maimonic system rather than a collection of remarks.

This vagueness, as almost all students of Maimon have emphasized, detracts not at all from Maimon's crucial role in the formation of post-Kantian idealism. It has been said by many scholars of the first rank, such as Bergman, Atlas, Cassirer, and Beiser, that the system of Fichte, and even those of Shelling and Hegel, are quite unimaginable without the pioneering strides of Maimon. Thus we learn that Maimon's ideas formed the bridge to these systems; yet when we approach the system itself we do not gain much beyond this historical insight.

It should be stated at the outset that the *Versuch* was not written as a composition with a purpose known to the author in advance, such that all the latter had to worry about is how best to deliver it to his readers. The book originated from remarks Maimon made about Kant's *Critique of Pure Reason*. This indeed is how he explains the genesis of his book with hindsight:

> The method by which I studied this work was quite peculiar. On the first perusal I obtained a vague idea of each section. This I endeavored to make distinct through my own reflection, and thus to penetrate into the author's meaning. This is properly called thinking oneself into a system. But as I had already mastered in this way the systems of Spinoza, Hume and Leibniz, it was natural that I should think of creating my own Koalitionsystem. This in fact I found, and put it gradually in writing in the form of explanatory observations on the *Critique of Pure Reason,* just as this system unfolded itself to my mind. Such was the origin of my book on Transcendental Philosophy wherein I develop each of the aforesaid system in such a manner that the unifying point common to all clearly emerges. (*Autobiography,* 143)

We also learn from this that Maimon supposed he had a system, and he explains that his system is the result of his original reading—*thinking oneself into the system*—of the *Critique.* This mode of contemplation had been previously applied to Leibniz, Spinoza, and Hume.

Maimon translates Kant into the terms of the problematic that is alive in his mind having read those philosophers. As Rotenstreich (1967, 534) pointed out, Maimon synthesizes Leibniz, Hume, and Spinoza with Kant, whose own thought forms a synthesis of Leibniz and Hume. Next, it is clear from this passage that the *Versuch* is made up of the remarks aimed specifically at the *Critique.* In point of fact, it would be wrong to surmise that Maimon put together an orderly system based on his notes on Kant. The *Versuch* is rather predominantly a reproduction of the very notes, only a subset of which concern the independent philosophical position he arrived at. For that matter, the *Versuch* is an exegetical commentary on Kant, analogous to Rashi's commentary on the Bible, or, even better, akin to the *Tosafot* commentary on the Talmud, in which Maimon was well versed. That was the literary genre in which he was reared, and the style in which he wrote his book on Maimonides *Givat HaMoreh.* If one reads the *Versuch* with this in mind, it is not difficult to see why his readers could not perceive any plan, while Maimon was satisfied he had a book. He was writing within a tradition where there was truth in all philosophies, where the texts added up to one whole, and the point was to grasp how it all cohered.

On a second reading of the *Versuch,* however, it transpires that Maimon was aware of this problem as besetting his writing, and he understood that the work could only be amended by a reworking; and in the preface he also explains that the composition is "an attempt that I intend to completely afresh in the future" (*Versuch* 11). Hence, having reread the manuscript he decided to assist the reader in a number of ways to grasp what is essential in his remarks. He writes a short survey, he adds a summary, and he sketches the outline of his plan.

An illuminating analogy that serves as a key character of his writing is suggested by something he says about his drawing abilities. "If I were instructed in this," he explains, "I should surely be a great painter, but not an accurate one, that is I would easily draw the main outlines of the picture, except that I would lack the patience to work it out precisely" (*Autobiography* 69). The survey and summary, then, facilitate the reading, but only insofar as they tell us what Maimon thought. They deliver his main ideas, however, in a language quite private to Maimon. Nevertheless, they help us locate the interpretive obstacle in reading him. Referring back to the drawing analogy, once we have these summaries, all that is required is the patience for precisely working out the details of his system.

It is important to emphasize Maimon's paramount concern, and the main difficulty in understanding him. Maimon, as is implied by the remarks in which he summarizes his main ideas in the *Versuch*, aspires to meet the challenge set by Kant in the *Critique* regarding the question of the possibility of metaphysics. He seeks to restore metaphysics' lost dignity, not by constructing a new metaphysics, but by a careful attempt to prove that, despite Kant's arguments, metaphysics is not impossible. For in Maimon's view, Kant tried to prove that the metaphysical urge is pointless, or at least should be restrained. Maimon was not trying to work out a full-fledged system; he was trying only to establish a narrower contention—namely, that it is legitimate to ask metaphysical questions. According to Maimon, Kant did not eliminate the possibility of finding a story that would lead us to the intelligible something underlying the phenomena.

As Maimon sees it, the history of metaphysics abounds in proposals and attempts to overcome various dualisms. Initially we separate the body from the mind, distinguish between the intelligible and the ephemeral, between form and content, and, after Kant, between understanding and sensibility. But these bifurcations are matched by the inverse problem of bridging the gap between the two aspects of being, or of cognition. From Maimon's viewpoint, Plato, Spinoza, the kabbala, and Leibniz are all concerned with different manifestations of one problem. Now this problem reappears in Kant's transcendental idealism as well. Against this background one can understand the centrality of the *quid juris* question, which recurs each time Maimon summarizes his main points in the *Versuch*, or when he outlines a work plan. By its very relevance, this question brings out that Kant's doctrine has not yet solved the classical philosophical questions, since these reemerge in the shape of the dualism of concept and intuition.

Maimon then attempts to show that it is possible to answer this question, at least in the field of mathematics, and this solution opens up the possibility of metaphysics in which the dualism between sensibility and understanding and the gap between phenomena and things-in-themselves both disappear.

Maimon's original gesture in the *Versuch* is, then, to extend the quid juris to the realm of mathematics and to declare that the problem of the link between thought and the sensible already arises there. This move never received due appreciation, although it does appear in Maimon's chart of his work plan, in that mathematical cognition is two tiered. Though both are a priori, the one points toward intuition while the other does not (without being merely logical) and we cannot comprehend the link between them. The only way out of the problem of the nature of the connection between these tiers is to suppose that the intuitive is no other than the intelligible itself in potential form: it is our limitations that lead us to grasp the intellectual pictorially.

Unlike Leibniz and subsequently Frege, Maimon does not regard this picturing as presupposing realism; rather, it is to be understood as implied by a reduction of intuition to its elements that occur in time. Kant held a static view of cognition, hence the dualism of concept and intuition seemed to him absolute. Had he taken into account that cognition could be dynamic, he would not have repudiated metaphysics.

The originality of Maimon's construction of what it is to be "external to consciousness" is remarkable. He bases it on his analysis of intuition. Like Berkeley and Kant, and for the most part like Hume and Leibniz, Maimon rejects the idea of a material object outside consciousness underlying the phenomena. He proposes that we analyze the awareness of something external in terms of reason, which is experienced by the subject as an entity being out there. The materialist-realist out there is in Maimon's view a projection originating in our ignorance of the conceptual foundations of intuitions, and it should be noted that this very explication of outsideness is based on the idea that intuition has an intellectual foundation. Hence, while in Leibniz (and even in Berkeley) it is only God's infinite consciousness that makes possible the world of objects, in Maimon it is rather our finitude, our own cognitive limitation, that enables consciousness of a world independent of us.

This idea is intelligible provided the solution to Maimon's quid juris question can be extended beyond mathematical cognition. The problems pertaining to such an extension derive from Hume's skepticism, that as conceived here challenges the very possibility of starting out from a fact whose conceptual structure we are yet to expose. Humean skepticism is thus a premise of Maimon's position. This is confirmed by Maimon's solution to the mathematical quid juris problem. It is the possibility of metaphysics that enables us to understand the significance of the possibility of skepticism. It turns out, then, that Maimon's critique of Kant concerns the central conclusion of the *Critique* insofar as the latter can be characterized as Kant's reaction to being waked by Hume from his dogmatic slumber. The series of philosophical influences on Maimon, that of Leibniz, Spinoza, Hume, and Kant, lead him to a tension, to a dilemma rather than to a synthesis. This description is consistent with the concise summaries of his system, like the one at the end of Maimon's preface to the *Versuch*:

> In particular I present to the reflective reader for inspection the following remarks. First, the difference between a priori cognition per se and pure cognition and the unsolved difficulty regarding the latter. Second, my own method of deriving synthetic judgments from the imperfection of our cognition. Third, the doubt concerning the quid facti question, regarding which, as it seems to me, Hume's objection is unanswerable. Fourth, I have hinted at the solution of the quid juris question and at the explanation of the possibility of metaphysics in general through reducing intuitions to their elements (reduzieren der Anschauungen auf ihre Elemente) which I call the ideas of reason. (*Versuch* 9)

The importance of this citation, which has not received due notice, is immense, if only one considers the opacity of the *Versuch*. For it allows for an initial ordering of the multiplicity of ideas in the *Versuch* and for a singling out of the ideas presupposed by the moves that give rise to this question. Abstract though it is, this description enables us to define what the exegetical problem ought to be: To explain the distinction between pure and merely a priori cognition, to elucidate the quid juris question, and to explain what is meant by "reducing intuitions to their elements."

But now, when we place these problems side by side, it is easy to see how they are connected to the philosophy of mathematics. Maimon's version of the quid juris question is no other than the question of the connection between a priori cognition modeled by geometry and pure cognition modeled by arithmetic. The "reduction of intuition to its elements" refers to a specific link between a priori cognition and pure a priori cognition that is required for the solution of the quid juris question and for the rehabilitation of metaphysics. Hence, although Maimon does not explicitly mention the importance of the philosophy of mathematics in the cited remarks, it is readily understood that this is a fundamental level, as it were, from which Maimon purports to examine the Kantian doctrine.

The picture that emerges thus is different from the usual in Maimonic studies. The topics of customary concern, such as the differentials, the criticism of the thing in itself, the fusion or confusion of skepticism and rationalism and other ideas undoubtedly preoccupied Maimon, but my main contention is that one's approach to these issues ought to develop out of the criticism of Kant's dualism, which occurs already in the analysis of the possibility of geometry, and Maimon's detailed solution thereof. If one reaches these topics from geometry, it is possible to show the unity of Maimon's thought, to point at the origin of his central moves, and even to find a way out of several apparent contradictions in his theories. Thus, the significance is clarified of his rationalistic idealism, as well as of his critique of the thing-in-itself, of Maimon's way of conceiving spontaneity, of his monism, of the meaning of the infinite intellect, and of the objection to the transcendental deduction. All these are illuminated only after having dealt with the intuition-concept dualism in mathematics. In particular, if this program is fleshed out one can also appreciate Maimon's

Kantian criticism and thereby be able to regard him as an independent philosopher, and not merely as an important bridge between Kant and German idealism.

Even the scholarly work that has noted the importance of the quid juris question has not appreciated Maimon's raising it in connection with mathematics. There has, moreover, been no attempt to explicate Maimon's solution, which was formulated in terms of "reducing intuition to its elements." It has usually been said that on the basis of differentials and the ideas of understanding that underlie intuitions one can resolve the quid juris question. But in the absence of an explication of the notion of underlying intuition, and in view of the obscurity of Maimon's notion of differentials (as he himself concedes and as is agreed upon by scholars), this type of description is notoriously prone to circularity: the differentials come to be *defined* as the answer to the quid juris question.

The first chapter in this volume provides the background from the *Critique* necessary for an understanding of Maimon's philosophy from the perspective offered here. I trace the dualism between a priori intuition and concept, between phenomenon and noumenon, and present them as deriving from Kant's analysis of the possibility of synthetic a priori cognition.

The second chapter is devoted to explicating the quid juris question as it arises in mathematics while explaining the distinction between the various tiers of mathematical cognition according to Maimon.

In the third chapter I formulate Maimon's answer, which is given in terms of the metaphor that intuition is an incomplete intellect through a special kind of reduction of a priori intuition to pure cognition. This kind of reduction I have dubbed *Maimonic reduction.*

The fourth chapter clarifies the implications of Maimon's answer to the quid juris question regarding the notion of objectivity. The connection is readily seen, since the need for sensuous intuition (which is opposed to intellectual intuition) implied by Kant's answer to the question about the possibility of a priori knowledge entails an abatement of the degree of objectivity we can ascribe to our cognition.

The monism entailed by the Maimonic reduction enables Maimon to rehabilitate Leibnizian dogmatism, and insofar as his monism does not extend beyond mathematics, one can also use it to establish skepticism.

We thus see the root of the Maimonic polarity: wherever Kant attempted to combine empiricism and rationalism, Maimon reinstates strict rationalism and thereby strict skepticism.

The fifth chapter illustrates this polarity in the case of space and time. Space is an intuition as well as a concept; the relation between space and its underlying concept is required according to Maimon to be identical to the relation between a priori cognition and pure cognition as explained in chapter 3.

Given Maimon's monism, it becomes possible to explain—in the sixth chapter—the concept of affection: to be affected is nothing but ignorance of the intelligible element underlying the sensible component of our cognition. Hence the idea of a thing as it is in itself as correlative to sensibility is seen to be superfluous. In fact, this analysis can be applied generally to every Kantian concept suggestive of existence outside consciousness. This gesture culminates in proposing the concept of consciousness *(Bewusstsein)* as primitive, and by defining Kant's concept of representation *(Vorstellung)* in terms that are internal to consciousness. By noting these moves we come to understand Maimon's claim about the intuitive intellect, of which the human intellect only partakes or is a schema of. This view of the intuitive intellect and of human consciousness renders superfluous Kant's transcendental deduction that presupposes that the "I think" does not bring about the manifold but only arranges it.

In concluding the argument, I define Maimon's relation to rationalism, idealism, and skepticism. Maimon's system is a synthesis between rationalism and idealism. Nevertheless, this system is a mere possibility, for Humean skepticism is still lurking in the background. This skepticism, however, is not wedged in the gap between the intuitive and the intellectual, and it is not predicated on the inference that seeking the intelligible element is an unending task, but rather based on the very question whether there is something underlying the world that appears to us. This tension between the possibility of metaphysics and the possibility of skepticism is not affected by two independent moves. Rather the skeptical side is constituted through the possibility of metaphysics.

Previous commentary on Maimon has emphasized his ideas on the category of substance and causality as well as his theory of differentials and the law of determinability. In chapter 7, therefore, I shall address these issues and place them within the context of the argument of this book. As in the case of space and time, it appears that Maimon advances his thesis on substance and causality as mere possibilities in order to show there is hope for a metaphysical account of such issues.

In his *Autobiography,* Maimon describes his philosophical development as a direct line from that of Hume, Leibniz, Kant, and Maimonides; however, this description tells us nothing about the place of the kabbala in his system. One should not forget that Maimon's first book, *Heshek Shelomo,* was a kabbalistic work, wherein he attempted to discover a synthesis between kabbala and philosophy. A deeper reading will reveal the important influence of kabbalistic ideas, including the importation of structures from Lurianic kabbala, as well as from the Cordoveros explication of the Zohar. In the epilogue I describe this kabbalistic dimension of his philosophy, which should not be neglected in any complete presentation of Maimon's thought.

1

The Kantian Challenge

In the attic of an old woman's house, funded by the help of his friend Ben David, Maimon pores over the ubiquitously discussed *Critique of Pure Reason*. The fruit of Maimon's labors is his composition about transcendental philosophy, in which he proposes a unified system comprising the doctrines of Spinoza, Leibniz, and Hume. The point of departure of Maimon's coalition of systems is described in his autobiography when he views the *Versuch* from afar: "The most important problem dealt with in the Critique, namely the 'quid juris?', I discussed here in detail in a much broader sense than conceived by Kant, and thereby I make room for Hume's skepticism in its full force. On the other hand, however, the complete solution to this problem necessarily brings with it the dogmatism of Spinoza and Leibniz" (*Autobiography* 223).

Contrary to many of his contemporaries, Maimon does not proceed to construct his system through the critique of the concept of the thing-in-itself. Rather, the return to Spinoza, Leibniz, and Hume is the consequence of the extension of the Kantian question of *quid juris*. The difference between these two ways of criticizing Kant becomes clear when we recall the two dualities that permeate Kant's thought: The dualism of phenomena and the thing-in-itself, and the dualism of intuition and concept. If we want to characterize Maimon's theory with regard to its *consequences,* then we ought to use the terms of thing-in-itself, affection, etc. Kant's criticism of the possibility of metaphysics of the type espoused by Leibniz and Spinoza is that they thought human reason could transcend experience and they considered things-in-themselves. However, the *logical* order of things is to first put forth the *quid juris* question—that is, Maimon's criticism of Kant's dualism of sensibility and understanding or of intuition and concept.[1]

When we note this fact, we understand Maimon trod a far more difficult path. The problematic of the thing-in-itself and the idea of the thing-in-itself affecting our consciousness are presumably problems that are more easily understood, while the problem of the dualism of concept and intuition is seemingly solved in Kant's system. Kant does indeed connect these two elements in his well-known remark that concepts without intuitions are void and intuitions without concepts are blind. However, this should not lead us to think that there is no intrinsic difference between the two classes. As Kant says: "The understanding can intuit nothing, the senses can think nothing. Only through their union can knowledge arise. But that is no reason for confounding the contribution of either with that of the other; rather it is a strong reason for carefully separating and distinguishing the one from the other" (*Critique* A-51, B-75). It would seem Kant required that the two components interact with one another, but nonetheless he distinguished between the two. I shall refer to this idea as Kant's interactive dualism.

In order to understand Maimon's quid juris we have to grasp both Kantian dualisms and the way in which they arise. The characterizing of intuition as a singular representation, as opposed to concept, which is a nonsingular representation, is not enough to make this into a problem. Why should it be more of a problem to connect a singular representation to a nonsingular representation than to connect one nonsingular representation to another? And the same is true of the characterization of concepts as mediate representations, contrary to intuition, which is immediate. Likewise, it won't do to say that Kantian intuitions and perceptions are conceptual by virtue of being different from subjective sensations, and to later try to connect them to concepts. This would not solve the quid juris.

For, in any case, it implicitly assumes the legitimacy of the connection between intuition concepts and sensibility already at the level of intuitions.

The Ideality Thesis

The representations of space and time are not an abstraction from experience, and they are not concepts that can apply to more than one object. They are purely a priori intuitions. This characterization includes components whose connection would be difficult to grasp if we were not dealing with Kant's theory: Although they are intuitions, they are not empirical; instead, they are a priori, yet not like other a priori concepts, such as causality and the concept of object. Furthermore, these properties are insufficient to substantiate what is most essential about space and time, namely that they are ideal. The ideality argument adds further that they are not things-in-themselves and they do not represent things-in-themselves.

Indeed, in order to argue that space and time are ideal, it is not enough to argue that space and time are intuitions, nor is it enough to argue that any object that is experienced must be experienced through these forms. We know this from comparing Kant with Leibniz, who held a similar view of space and time. Leibniz also claimed that the body (as opposed to the monads) indeed exists in space and time, but therefore it belongs to the world of phenomena. For Leibniz, space is a well-founded phenomenon in the same way the rainbow is a phenomenon. The ideality of space and time is inextricably bound up with the claim that they are not things-in-themselves, nor are they *grounded* in the things-in-themselves. Leibniz argued in favor of some kind of correspondence, which may be understood in terms of an isomorphism or in terms of the possibility of translating space-time features into attributes of substances. Kant, as we shall see, criticizes him on this score when he says, "Space does not represent any property (Eigenshaft) of things-in-themselves, nor does it represent them in their relation to one another" (*Critique* A-26, B-42).

It should be remembered that the thing-in-itself is not defined as a world we are unfamiliar with but as a complement of sensibility that is required in order to regard sensibility as somehow derived from what is not sensuous. Whenever we see a tree or a clock we know, according to Kant, we are affected by a thing-in-itself. However, this characterization actually invites the linking of phenomena to the thing-in-itself rather than totally discouraging it. In modern science, atoms and magnetic fields are not phenomenal, yet we study them by observing phenomena. In point of fact, most epistemologies prior to Kant's, with the exception, perhaps, of radically skeptical or Parmenidian approaches, held that what is beyond phenomena can be known through phenomena. It is the ideality thesis that blocks our access to the world of things-in-themselves: by studying phenomena we cannot learn anything about what is beyond them. Kant, like many realists, does not consider space as a picture of a nonspatial reality. Yet by adding the ideality thesis Kant restricts the spatial mode of experience to the human sphere. What, then, is the reason to think that if we have synthetic knowledge of space we are compelled to concede that space is a mode of representation confined to humankind and does not reflect any property of things-in-themselves?[2]

The starting point for an answer or a description of the transition from synthetic knowledge to the ideality of space and time can be found in the following short argument, which is based on the apodictic nature of geometric knowledge.

We know generally valid things about bodies and the relations between them with certainty. For example, we know for certain that it is impossible to draw four lines all issuing from a single point and all perpendicular to one another. If geometry were the study of things-in-themselves, we would not be able to explain this certain knowledge. This conclusion follows from the premise that if these properties were of things-in-themselves, we would not be able to know for certain that which we know; there would always be room for error. This premise could also be expressed as follows: adopting a realistic stance toward space and time would allow for a possible disparity between the truth and my own beliefs, and would thus leave room for error regarding geometric axioms. But there is no possibility of doubt regarding geometrical axioms, and it is thus impossible to adopt a realistic stance toward space.

This point can be illustrated as follows: The fact that I cannot imagine four perpendicular straight lines all issuing from the same point does not lead to the conclusion that there are in fact no such lines. What is more, I imagine lines and circles, which are governed by geometric truths. If these lines and circles were the representations of something beyond my powers of imagination, there would be room for doubt regarding geometric truths. For my imaginative powers, in this case, pose a subjective limitation, and there is always the logical possibility that reality is one way and my subjective image of it another. Because I know the laws of geometry constitute a science, which provides us with objective knowledge, I have no alternative, so Kant argues, but to conclude that the laws of geometry do not refer to a reality beyond that which is accessible to me. It is important, however, to emphasize that without categories there can be no talk of substances, so geometry cannot tell us about reality. But for the foregoing argument it is sufficient that the principles of geometry constrain the properties of substances. More precisely, they are not either true or false by virtue of properties that would persist even independently of the fact of their being apprehended by humans, and in fact they describe my representational faculty.

There are two features of geometric axioms at work here that reinforce Kant's demand for an explanation: the axioms are general and yet synthetic. For how am I to explain that I know something general about a certain field before ever having had any experience of it? True, I understand the concepts employed in the argument in question, but my knowledge about the relation between the premise and its conclusion goes beyond anything I could know simply by virtue of understanding these concepts. For although I cannot comprehend the possibility of someone subscribing to both a statement and its contradictory, I can fully comprehend the possibility of four-dimensional or even infinitely dimensional space. The possibility of multidimensional mathematics and the impossibility of mathematics that does not presuppose the law of contradiction both lend support to Kant, as is stressed now and again in commentaries. This leads me to ask how it is that I know that any four given lines will obey the law in question?

The force of this short argument lies in the fact that it allows us to make the leap from synthetic a priori statements to the fact that we are not dealing with properties that exceed the realm of phenomena. Although this is not exactly the way the account appears in Kant's work, it seems to be essentially what he had in mind. One can find some of the stages of this argument in the "Transcendental Exposition" in the *Critique* (A-25, B-41), where the claim that geometrical axioms do not refer to things-in-themselves is presented as a conclusion from the character of geometrical axioms. In the absence of any other obvious account, one can reasonably assume that it was an argument much like this one that Kant had in mind when he remarked, in a letter to Reinhold, that the fact that geometry involves synthetic statements is sufficient to establish that these statements do not apply to anything beyond phenomena (Kant to Reinhold, 12 May 1789, *Letters* 136–43).

Yet the problem with this argument is related to its abstract nature. By this short route, we can prove that the laws of geometry are not valid beyond that which we can perceive, without committing ourselves to any position regarding the nature of space and time. The above argument demonstrates that geometric axioms do not describe things-in-themselves, but one cannot immediately see how this leads to the conclusion that space and time are ideal. Kant himself confirms this point in the above-mentioned letter to Reinhold. He states that "the existence of synthetic a priori statements is, in itself, sufficient to establish that these statements do not apply to anything beyond the phenomenal." He goes on to explain: "this fact can be established even if we do not know that space and time are the forms of our sensibility." How, then, can one demonstrate that space is indeed a form of our sensibility?

First, one must supplement this discussion with the "Metaphysical Exposition," which we have hitherto neglected. The scope of the present work prevents me from providing a detailed discussion of the arguments of this inquiry, which have been analyzed in detail in several commentaries on Kant. Suffice it to mention that, in the course of the inquiry, Kant provides a logical-phenomenological analysis of space and time. This section proves, or at least argues, that we locate everything that appears in space in a single space. The event of my eating ice cream yesterday, the First World War, and my act of writing today all took place in the same space. This leads Kant to conclude that space is not a concept but an intuition. If *space* were the name of a concept, I would be capable of thinking in several different spaces, and of doubting the location of all events in the same single space. Here, too, the fact that we cannot entertain such doubts leads us to conclude that space is an intuition. But as soon as we admit that space is an intuition, we are forced to add that it is a special type of intuition. There is a categorical difference between the intuition of space and other intuitions that appear in space and that are conditioned by space by presupposing it as a background. Kant concludes from this point that the representation of space is not something derived from empirical experience, but rather itself constitutes a condition for anything that can be experienced. These remarks precede the above-proposed analysis of geometry, and they show us that space is an a priori intuition.

Thus we cannot assume that the laws of geometry apply to things-in-themselves (without forfeiting their apodictic character), and we know that the representation of space is a priori and that it is a condition for anything appearing (in space). But we have to be cautious here: the first qualification cannot be formulated as the claim that we know that the laws of geometry do not apply to the world of things-in-themselves. We cannot say that we know that things-in-themselves do not conform to Euclidean geometry. All we know is that we have no reason to assume that geometrical axioms are valid in the world of things-in-themselves, and that the question as to the relation between our world and the world of things-in-themselves is not a question for science. We know that what makes geometry true is not the world of things-in-themselves, but we have no way of knowing that geometry is not true in that world. Apparently, if we knew, apodictically, that geometry did not apply to the world of things-in-themselves, we would have to make this world an ideal world. But neither the short argument described above nor the properties of space as an a priori intuition can provide us with the whole picture about geometry, space and time, and the transcendental exposition; we cannot supply this picture without accounting for the Kantian concept of affection. The concept of affection is surely one of the most significant concepts in the Kantian system, as well as one of its vaguest notions.

As we shall see in the following chapters, Maimon reverts to this concept, and therefore we cannot ignore it in interpreting Maimon. But then its importance for understanding is clear from Kant's own contention that space, like time, is a general mode in which we are affected. This remark incorporates a rich and essentially positive conception of the ideality of space, a conception we cannot ignore even in this short introduction. If we wish to understand the meaning Maimon attaches to this notion at a later stage, then we must first offer a few remarks, however general, regarding Kant's use of it.

The concept of affection first appears in the *Critique* in the "Transcendental Aesthetic," in the course of the discussion of the distinction between concepts and intuitions. "But an intuition takes place only in so far as the object is given to us. This, again is only possible, to man at least, in so far as the mind is affected in a certain way. The capacity for receiving representations (receptivity) through the mode in which we are affected by objects, is called *sensibility*. Objects are given to us by means of sensibility, and it alone yields intuitions" (*Critique* A-19, B-33). The concept of affection thus connects the concept of the object with the distinction between intellect and sensibility. We are affected by objects, and we respond to this with sensations. At first glance, this passage may give the impression that such affection is something like the stream of photons hitting the retina, to which the mind responds in the form of a sensation. But viewing the world as a collection of atoms, and picturing electromagnetic fields as located in an objective space and influencing our sensory organs, are quite foreign to Kant's conception and opposite to his general conclusion.

The idea that we are affected in sensation is associated with the idea that the active element in cognition is the intellect (and not the photons). But this may lead us to think, not without reason, that the spontaneity of the understanding and the passivity of sensation consist in the fact that while we can think of anything we wish, when it comes to sensation, we are affected because we are forced to sense things. When I open my eyes I am not free not to see, nor am I even free to determine what I see. But in the intellectual domain of concepts, I can think of anything I please.

This interpretation, endorsed by Fichte and prevalent ever since, will not do.[3] For there is no doubt that in order to understand the active aspect of the intellect one has to consider the Aristotelian concepts Kant inherited through philosophical tradition. The intellect, like the Aristotelian form and the "Active Intellect" is the active agent for Aristotle. This is what makes anything what it is; it is the shaping agent conferring unity on plurality or making the many into one. For Kant, the understanding and the concept are active by virtue of their power to connect elements and to group them into a unity. Like the Aristotelian form, it is the concept that makes the given thing what it is. To be shaped as something—a circle, a lion, or a table—means to be related to other representations in a certain way. Without the concept of lion we would have a diversity of unconnected images. The concept connects by its very shaping activity.[4]

The passivity of the sensory data is, then, connected to several ideas. First, it is passive in the sense that I am not the source of my representations, which are, rather, imposed on me. Secondly, it is passive in the sense that the mind molds the sensory data, thus causing them to be perceived in a particular way. These ideas are somewhat vague, but they form the basis of Kant's argument when he explains the ideality of space and time. Space is not an empirical representation derived from experience, and it is the condition for all phenomena. At the same time, the theory of the ideality of space and time amounts to the claim that these are general ways in which the subject is affected or, as Kant says, forms sensibility. This claim takes us much further than the claims that space is a priori (in other words, not derived from experience) and that it is not a concept but an intuition.

However, we are now faced with a further difficulty. Even granted that we know geometry is not founded on the character of things-in-themselves and that people are affected by sensation itself, we must still consider what lends support to the claim that space is merely a particular way in which we are affected. How do we know that there is such a thing as a general way of being affected?

The question is not how we know that our experiences are always in space and time—to this question we will return when discussing Maimon's theory of space. Sensory data appear in space and time, and one could say that space affords the stage for the senses. The question is rather how we can know this stage is just the way in which the subject is affected. In more concrete, although somewhat simplistic terms, one could argue that a fish in the sea knows for certain that its experiences are always in water, but no fish is warranted in concluding from this that the sea only exists from its singular point of view. On this point, the Kantian thought experiment that abstracts all the components of sensation, leaving us with the form of the sensation, will not suffice as an answer either. Let's assume that I abstract everything material from intuition; how do I know that what I shall finally reach will be a general mode in which the self is affected? What is it that guarantees in advance that the abstraction will not yield pure extension, as did Descartes' abstraction of all sensible qualities from his wax?

In order to find an answer, or at least a useful clarification, we must conjoin all the ideas presented thus far:

1. We concluded from the short argument that the range of application of synthetic judgments must be restricted to that which is accessible to us. It is not grounded in the realm of things themselves.
2. The phenomenological exposition, which appears as part of the metaphysical exposition, implies that space and time is the background to all our images.
3. Kant accepts the traditional postulate that in sensation we are affected and that an object is possible for us only if it affects our powers of imagination.

If we put all these points together, we can clearly see that space and time are only forms of sensation dependent on the subject—that is, they are the general modes in which we are affected. This argument cannot be logically formalized since we do not possess precise definitions of the terms Kant deploys. Nevertheless, the combination of these three points recommends and clarifies the claim as to the ideality of space.

Kant takes this argument even further. Not only is space dependent on the mind in some abstract sense, but it also cleaves to the mind. Kant stretches further the Cartesian idea that the mind is easier to know than anything else and argues that if anything can be known, it must belong to the mind. In other words, Kantian idealism is not founded merely on the impossibility of my being able to claim that I cannot imagine a contrary state of affairs, while in reality things may well be different (as explained above). The claim as to the ideality of space is also based on positive considerations: the subject knows something which he, as subject, contains. As Kant explains: "How, then, can there exist in the mind an outer intuition which precedes the objects themselves, and in which the concept of these objects can be determined *a priori*, exist in the human mind? Manifestly, not otherwise than in so far as the intuition has its seat in the subject only, as the formal character of the subject, in virtue of which, in being affected by objects, it obtains immediate representation, that is intuitions, of them" (*Critique* A-25, B-41). Space is a secondary property, analogical in ontological status to the sense of color. As Kant says and repeats, this secondary property does not stray from objects to enter the mind. It is a modification of the subject: We do not exist in space and time; rather, we carry them within us.

Let's leave this strand of the discussion aside for a moment and illustrate this point. The space surrounding us, whose dimensions have inspired wonder throughout the generations, and which contains the planets that filled Kant with awe, is a mode, a determination, of the inquiring subject. It is difficult to refrain from a comparison with Spinoza's system in this context. Spinoza also regarded space as an attribute dependent on the subject, but the subject in his case is God, and the infinity of space is an expression of God's infinitude. Kant rejects this theocentric position but retains the idea that space is carried by the cognizant subject. Just as Spinoza's space is dependent—without God there is no space—so too is Kant's space dependent; but God is replaced by Kant with man. A comparison with Newton leads to a similar conclusion. Newton saw space as God's perceptual organ. Space, for Kant, is also associated with perception, but it is man's perception. Removing God from the picture is not a postulate of the Kantian system but is the consequence of shifting the discussion from ontology to epistemology. This also constitutes a shift away from metaphysics, with its pretension of knowing things-in-themselves, to metaphysics in the transcendental, critical sense.

Kant's Interactive Dualism

Kant's doctrine is entirely based on the duality of understanding and sensibility, or of concept and intuition. What Kant primarily has to say in support of this distinction is the rejection of Leibniz's view that sensibility is a confused concept. In fact, the ideality of space and time is only one of the conclusions one can draw from the fact of the existence of synthetic a priori judgments together with certain other premises, but it does not provide a complete explanation of the possibility of synthetic a priori knowledge. Surprising as it may seem, there is no such explanation in Kant's "Aesthetics." As Kant explicitly informs us: "And finally, Mr.

Eberhard says on page 316: 'One seeks in vain for Kant's principle for synthetic judgments.' But that principle is unequivocally presented in the whole Critique, from the chapter on the schematism on, though not in a specific formula. It is this: all synthetic judgments of theoretical cognition are possible only by the relating of a given concept to an intuition. If the synthetic judgment is an experiential judgment, the intuition must be empirical and if the judgment is a priori synthetic, there must be a pure intuition to ground it" (Kant to K. L. Reinhold, 12 May 1789, *Letters* 141).

Kant understands the need to go beyond concepts in order to establish synthetic judgments in a positive sense. In explaining, in the *Critique*, the principle of all synthetic judgments, he states: "Granted, then, that we must go out beyond a given conception in order to compare it synthetically with another, a third something is necessary, as that wherein alone the synthesis of two concepts can be achieved. What now is this third something that is to be the medium of all synthetic judgments? It is only a whole in which all our representations are contained, namely, inner sense to it, and its form a priori, time. The synthesis of presentations rests on imagination; and their synthetic unity, which is required for judgment, on the unity of apperception. In this, therefore, we must look for the possibility of synthetic judgments" (*Critique* A-155, B-194).

These passages clearly demonstrate the connection between the explanation of the possibility of synthetic cognition and "correspondence between intuition and concepts," or the reference to "a third element which is to be a medium." Whether we argue for the correspondence between intuitions and concepts—examples of which are provided by geometry—or whether we decide that the intuition of time as a medium that enables us to establish a synthetic connection is more appropriate for an explanation of the possibility of principles such as causality, we will have to acknowledge an interaction between the conceptual and the intuitional.

From the above it can be seen that Kant resorts to the correspondence between concept and intuition, which is not a concept, as the only way, in his view, to explain the possibility of synthetic knowledge. Insofar as such a move is plausible, it affords the right conceptual background for an understanding of Kant's conception of geometrical *construction*, without which Kant's dualism is merely abstract. As will soon transpire, Kant does not borrow the meaning of construction from the mathematician nor from previous philosophers. He molds this concept through the problem of the fecundity of the mind.

To be able to appreciate the meaning of construction it is necessary to introduce, even if most briefly, the Kantian notion of a schema, since the linking of intuition to concept occurs through the mediation of the schema situated between understanding and sensibility. The schema, according to Kant, is not a specific image, as Berkeley thought, nor a general image, as Locke would have it. Furthermore, Kant is not claiming that the schema represents the concept—as Aristotle argued long before him—nor that it is an incomplete concept in need of determination, as Leibniz believed. The schema is an image of the *method* whereby a picture of a concept is drawn. As such it is related to that general framework through which alone specific pictures are possible, namely space.

We learned from "Aesthetics" that space is just the general manner in which we are affected. And the schema is the manner in which our sensibility is specifically affected, in accordance with a specific concept. This act of matching images to concepts allows Kant to respond to the question of the possibility of synthetic a priori judgments. Without the schema and the resulting interactive dualism, we would be faced with one of two options: Either we would have to regard geometric construction as a particular case or we would have to regard it as a merely heuristic means of looking at geometric concepts. In the first case, geometric inference would be like inductive inference; in the second case, we would end up with Leibniz's position, according to which geometric arguments are essentially logical. The claim that geometry is a collection of synthetic a priori judgments rejects both alternatives.

The above discussion about space enables us to appreciate the opening Kant found through which to explain the geometer's need for construction. The idea of the act of matching the image to the concept yields the following conception. At the time of providing a proof on the basis of a construction, we do not see the phenomenon that appears to the mind as a specific image with certain features. In construction, the concept is pictured or translated into an image, and what is seen by the geometer attempting the proof is the act of construction, not the constructed product. It is the *act* of realizing that is essential and not the realized picture.

When we focus on the mere final product, we see the construction as an individual image, and this constitutes an empirical intuition. In thus specifying the image, one can measure its angles. On the other hand, in order to construct and sketch a drawing one certainly needs to draw, but the focus of attention is on the act of drawing according to a concept or a rule, not the finished drawing itself. We let the process of construction go along in our imagination or on a piece of paper, and when we perceive the phenomenon in question, it is not yet certain whether it is an empirical or an a priori intuition. It is only when we determine the way in which we specify it—as an image or as the act of creating an image—that we can determine whether it is an empirical or an a priori intuition.

This also explains why it makes no difference in Kant's view whether the picture is drawn in our imagination or on a piece of paper; the main point is that what we follow through is the act of construction and not the final product of the construction—that is, the image. In following the act of construction, our attention is focused on the manner in which we come to perceive an image. Then, in perceiving the image as it comes into being, we are entitled to infer propositions, which are valid in a necessary, general, and nonanalytical sense. The following quotation confirms and summarizes this interpretation:

> To *construct* a concept means to exhibit *a priori* the intuition which corresponds to the concept. For the construction of a concept we therefore need a *non-empirical intuition.* The latter must, as intuition, be a *single* object, and yet none the less, as the construction of a concept (a universal representation), it must in its representation express universal validity for all the possible intuitions which fall under the same concept. Thus I construct a triangle, by representing the object which corresponds to this concept, either by imagination alone, in pure intuition, or in accordance therewith also on paper, in empirical intuition—in both cases completely *a priori,* without having borrowed the pattern from any experience. The single figure which we draw is empirical and yet it serves to express the concept without impairing its universality. For in this empirical intuition we consider the act whereby we construct the concept and abstract from the many determinations (for instance, the magnitude of sides and the angles) which are quite indifferent, as not altering the concept "triangle." (*Critique* A-712, B-742).

This position of Kant's is original. To the best of my knowledge, it has not been put forth by anyone else either before or after him.[5] It explains the geometrical construction in the broad context of the fecundity of the human mind and the nature of space and time.

We can now see that the dualisms in Kant's doctrine are intertwined: The ideality of space is connected to space as an intuition rather than a concept, whereas the interaction between intuition and concept is inextricable from the idea that space is a form of sensibility. I now turn to the general picture that emerges from the preceding discussion. Synthetic judgments are not derived from definitions or from empirical intuitions. They are possible due to the correspondence between images and concepts. Our minds grasp a concept by means of a definition, but it is only when it is realized in an image that its subjective dimensions—that is, those properties that exist only from a human point of view—can be perceived. That is because the correspondence between concept and image demands that we pass through the framework of space and time, which is the condition for anyone's being affected by images. The metaphor of spectacles is sometimes used to explain this state of affairs: space is like a pair of spectacles that separates us from the world of things. This metaphor can be improved upon by means of another image: space and time envelop us like a kind of big balloon (the form of sensation). Outside the balloon are concepts. When these concepts are pictured, they press against the surface of the balloon (when the schema effects the correspondence between an image and a concept). We look at concepts by looking at the protrusions they make on the balloon's surface ("it is impossible to think of a line unless it is drawn in the imagination"). However, this also includes reference to the surface of the balloon (the general capacity of the subject to be affected). The additional knowledge which we attain in this manner goes beyond what is true merely by virtue of the concept's meaning, and this is the source of synthetic a priori judgments.

We thus realize that when proving geometrical theories, the drawing we make is not merely a convenient aid to following the proof, or to understanding an argument that would have been valid independently of graphic representation. This could usefully be compared to the use of Venn diagrams in order to check the validity of logical syllogisms. In this case, looking at the picture is an easy way to ascertain something that has no essential relation to the picture or even to the possibility of drawing it. The logical validity of a syllogism is inseparable from rationality itself (and Kant would probably have referred to looking at Venn diagrams as an "analytical intuition a priori"). But the intuition involved when we prove a general, nonanalytical proposition is of a different kind, as it involves a subjective condition of our own, for the image is not a transparent straightforward reflection of the concept. This is the flip side of Kant's argument in "Aesthetic," to the effect that space and spatial relations do not reflect the intelligible world.

We can now refine the distinction between Kant and Leibniz. As stated, Leibniz did not think space had an independent existence. The monads are not bits of space, but they have a spatial position. Leibniz believed, however, that relations between phenomena in space are isomorphic to the intelligible essence associated with monadic properties, and that it is possible to make the transition from phenomena to their essence. On this point, Kant states as follows: "Doubtless, indeed, there are intelligible entities corresponding to the sensible entities; there may also be intelligible entities to which our sensible faculty of intuition has no relation whatsoever" (*Critique* A-253, B-309). And in elaborating his critique of Leibniz, he adds: "But even if we could by pure understanding say anything synthetic in regard to *things-in-themselves* (which, however, is impossible), it still could not be applied to appearances, which do not represent things-in-themselves" (*Critique* A-276, B-332).

There is no systematic connection between the world of science and the world of things-in-themselves, and, in any event, science is not the study of the connection between these two worlds. The image of the tree I see before me, the image of the house I see beside it, and the image of the other tree behind the one next to me all reach me in a manner of which I am partly passive. Furthermore, I am aware of the fact that beyond all this in a non-spatial sense, there are things unrelated to the spatiotemporal scheme. But I cannot conclude from this that the same reality lies behind both trees. Accordingly, the respective distances between the objects do not correspond to conceptual connections, which, as it were, form the grounds for these distances. But this does not entitle me to infer that the (nonphenomenal) "thing" underlying this tree and the one "behind" that tree share a common reality. Nor is there any nonphenomenal correlative to the one tree being closer than the other. If some mind could establish a connection to the world of things-in-themselves, it would have no way of knowing what was happening to me in the world investigated by the natural sciences. In contemporary terms, Kant is speaking of a kind of supervenience, of the world of phenomena on the world of things-in-themselves. In the theological terms in which Maimon, upon entering Germany, was well versed, we might say that a God who is unbound by the conditions of sensibility but connected to the realm of things-in-themselves is incapable of knowing the world of man.

§ 2 §

The *Quid Juris* Question

For some time before his discovery of Kant, Maimon could be described as philosophically blasé. A dozen years earlier he had exhausted his interest in kabbala; he was "eager to see knowledge in its natural light and not veiled by allegory and parable" (*Autobiography* x). Upon becoming acquainted with the philosophy of Wolff, he began shaking loose from the grip of Maimonides and even wrote in criticism of the latter's Thirteen Principles (except for the eleventh, concerning reward and punishment!). He immersed himself in Spinoza, with whose teachings he was already familiar, but, predominantly, he had devoted two years to the study of languages and fine arts. The latter at first struck him as odd, but gradually they came to amuse him.

One does not need to know much about Maimon's character to be able to guess just how enthusiastic and joyous his immediate reaction to the new theory would be. Even though his entire life was devoted to the quest for metaphysical truth and perfection, one should not think that Maimon meets the negative conclusions of the *Critique* with the paralysis of a von Kleist. The point where he decided to start his dialogue with Kant, which also led to his coalition of systems, was the *quid juris* question.

Maimon's quid juris question constitutes a demand to understand the connection between concepts and intuitions. It may seem that Kant's dualistic interactive thesis explains this connection. However, to believe this would be to ignore the distinction between the question of the very existence of the connection and the question of how to account for it. The quid juris question arises only because of the premise that form and content are indeed connected and operate in harmony. We assume, with Kant, that concepts without intuitions are void and intuitions without concepts are blind, but we want to understand how this connection is possible. In other words, we take the connection for a fact; that is, we answer the *quid facti* question in the affirmative, but we still want to understand the fact in question.

When directed against Kant this question may seem odd, for was it not Kant who first gave this question its name and solved it in his first *Critique*? In general, according to Kant, concepts should be classified into three types, on the basis of what and whether one is justified in employing them. These types are empirical concepts, such as gold and apple; usurped concepts, such as fortune or fate, employed with doubtful warrant; and concepts of understanding, with categories such as cause and substance, which are not grounded in experience. The following question arises regarding the last category: "For since empirical proofs do not suffice to justify this kind of employment, we are faced by the problem how these concepts can relate to objects which they yet do not obtain from any experience. The explanation of the manner in which concepts can thus relate *a priori* to objects I entitle their transcendental deduction" (*Critique* A-85, B-117).

Kant replies with his transcendental deduction and puts things to right in his answer in the chapter on schematism. Maimon does not accept Kant's answer to his question, nor does he accept the consequences of the possibility of the geometric intuition blocked by Kant's interactive dualism and by Kant's ideality of space and time. He chooses to engage Kant at the finishing line, and he raises the quid juris question, precisely in the light of Kant's answer as it is developed in his schematism.

In his schematism answer to the quid juris question, Kant assumed that the relation between empirical intuitions and pure concepts is problematic. Notwithstanding, he completely neglected the problem of the relation between pure concepts and a priori intuitions, already introduced in the preceding chapter to account for the synthetic character of geometrical knowledge and concepts. This indeed was the formulation of Maimon's question in his first letter to Kant.

Fortunately, we have an answer from Kant to Maimon's question, and the first part of this chapter studies Kant's response. To come to terms with this response Maimon had to sharpen his critique of Kant and transfer the quid juris question to the field of mathematics. This gives us the first contrast between Kant and Maimon, as the latter did not think there was any special relevant problem arising in the field of mathematics, whereas Maimon in his *Versuch* devoted enormous efforts precisely to examining the warrant for applying pure mathematical concepts to a priori intuitions. These efforts have not received due scholarly attention.

The Kantian Response

The first formulation of the quid juris question by Kant, as shown above, was the demand to explain in what way it is legitimate to apply pure concepts such as cause and substance to a posteriori objects. This formulation presumably poses the requirement of presenting objects for the pure concepts, which otherwise would be empty. Kant does sometimes postulate the condition that concepts be realized through experience, otherwise their meaning remains problematic. However, a more careful reading indicates that the question is not whether those concepts are realized in some object. Maimon grasps this point well, after differentiating between two kinds of possibility. A concept that does not fit any object is for both Kant and Maimon an empty concept. Actually applying it to an object explains why, for example, a concept like cause differs from an invalid concept like fate. But even having satisfied this demand, we are still left with the pivotal question of why Maimon calls it the explanation of the necessity of the possibility—namely, showing that the possibility of there being a relation between a category and experienced objects is not incidental. The best example of this is the mind-body problem. Even if we know that the mental and the physical work together in harmony, we may still inquire about the intelligibility of the connection between the two. Maimon does indeed generalize this question to include the mind-body problem (*Versuch* 62).

We have to be precise about Kant's wording. The question here is not about the connection between categories and objects, for the idea of an object is itself a category; thus to require that an object fit the category of objects is senseless. On second reading it seems that Kant's problem is how to explain the application of pure categories to experiential intuitions. We thus shift from inquiring about the relations between categories and objects to inquiring about the synthesis of categories with intuitions that produces the objects and a unified experiential world; this is made clear in the schematism. Kant asks: How can we explain the applicability of pure categories to the a posteriori material? His reply is as follows:

> Obviously there must be some third thing, which is homogeneous on the one hand with the category, and on the other hand with the appearance and which thus makes the application of the former to the latter possible. This mediating representation must be pure, void of all empirical content, and yet at the same time while it must in one respect be *intellectual*, it must in another be *sensible*... Now a transcendental determination of time is so far homogeneous with the category, which constitutes its unity in that it is universal, and rests upon an *a priori* rule. But, on the other hand, it is so far homogeneous with appearance, in that time is contained in every empirical representation of the manifold. Thus an application of the category to appearances becomes possible, by means of the transcendental determination of time, which, as the schema of the concepts of understanding, mediates the subsumption of the appearances under the category. (*Critique* A-139, B-178)

Kant solves the problem by means of a mediator, in a manner reminiscent of a puzzle in which one has to join two pieces with nonmatching ends and thus must search for a third piece, one end of which fits onto one piece and the opposite end of which fits onto the other. This missing part is none other than the sensuous a priori, which is revealed or designed in the "Aesthetics."

At this point, Maimon poses the following question: "Mr. Kant is trying to escape from this difficulty by assuming that time and space and their possible determinations are *a priori* representations within us, and therefore we may attach to a certain duration of time, which is *a priori*, the concept of necessity, which is also *a priori*. But since intuitions, as has been proved, even if they are *a priori* they are different from intellectual concepts, this assumption does not take us very far" (*Versuch* 65). Solving the problem by means of a mediator, as we have seen Kant do, is a typical move for the brand of dualism that involves interactive components. However, attached to this move is a typical problem: the connection between the two ends and the mediator linking them is not quite clear; we seem to be faced again with the classical third-person problem.

Now, the connection between an a posteriori phenomenon and the mode of sensation may be obvious. If space is the condition—the general background, as it were, for all phenomena—then every empirical fact presupposes space. The representation of a table sits clearly within the image of space, in the same way as a particular segment of time is anchored in the general flow of time in which all phenomena occur. But the nature of the connection between the pure concept and the sensuous a priori is not at all clear. The sensuous a priori is not determined by the conceptual understanding and is not a part of it. In other words, we cannot think of a spatiotemporal event without presupposing space and time, but we can conceptually comprehend the temporal succession of the faculty of the imagination without presupposing causality. Similarly, one can contemplate the concept of cause, which involves a necessary connection between things, without recourse to the mode of sensibility. How, then, are we to explain the synthesis between time and the categories?

Even if it is difficult to formulate this question in terms extraneous to the Kantian system, insofar as we are dealing with typical moves within the Kantian system, it would seem that the question is legitimate. Maimon could present Kant with a sharper formulation of the question, from within the system, namely: If you thought the relation between the category and the a posteriori phenomenon required a special explanation because they are not homogeneous, why do you not follow the same reasoning with regard to the relation between pure concept and a priori intuition?

This question reached Kant at the height of his career. Kant told Marcus Herz, who acted as mediator between him and Maimon, that since he was sixty-six years old while writing *The Critique of Judgement* he was inclined not to reply at all and simply return the big packet Maimon had sent him. Still, the significance of Maimon's remarks convinced him to find time to reply to them. Thus it would appear Kant was unable to devote the necessary time and attention to Maimon's question. Later, as well, in the letter to Reinhold mentioned in the introduction, he expresses the wish to relegate Maimon's *amendments* to treatment by others.

The first part of Kant's reply is that, without presupposing the interaction between the intellectual and the sensuous, we cannot comprehend the possibility of a priori synthetic judgments. One might be inclined to conclude from this dialogue that Kant's reply to Maimon amounts to admitting he had to make this presupposition in order to solve the problem of the possibility of knowledge. This would be like regarding Kant as a scientist adopting a counterintuitive hypothesis in order to explain observations that cannot be otherwise accounted for. And, indeed, were this Kant's response, it would constitute, at most, a response to the question regarding the *fact* of the connection between concept and intuition. However, Maimon's question is not about the existence of this connection but rather about how it is to be rationally explained, assuming it is a fact.

The second part of Kant's reply to Maimon is that his question cannot be solved by the human mind. "But we are absolutely unable to explain further how it is that a sensible intuition (such as space or time), the form of our sensibility, or such functions of the understanding as those out of which logic develops are possible; nor can we explain why it is that one form agrees with another in forming a possible cognition. For we should have to have still another manner of intuition than the one we have and another understanding with which to compare our own and with which everyone could perceive things in themselves. But we can only judge an understanding by means of our own understanding, and so it is, too, with all intuition. It is, however, entirely unnecessary to answer this question" (Kant to Marcus Herz, 26 May 1789, *Letters* 153).

Kant replies that in order to answer the question about the connection between concepts and intuitions we need a special kind of understanding and intuition, capable of perceiving the correspondence between our intellect and things-in-themselves, which in turn would oblige us to step outside ourselves and perceive things-in-themselves. But this reply to Maimon has a few obvious weak points: Kant is of the opinion that the reply to Maimon's quid juris question would necessarily involve referring to an intellect and intuitions distinct from our own. But he provides no argument as to why this is the only way to reply to the question. Kant criticizes what *he* claims is necessary to respond to the question, but how does he know that there is no other way to respond to the question about the correspondence between intuitions and concepts except through stepping outside ourselves?

Kant tries to force anyone purporting to solve this question into the absurd position of perceiving the thing-in-itself. But the commitment to the existence of things-in-themselves, not to mention the inability to perceive them, is an integral part of Kant's position. Invoking these views does not constitute a refutation of Maimon's question, which poses a challenge to our inability to know whatever it is that phenomena are founded on, and thus constitutes a challenge to the very foundations of the Kantian system. One could even go so far as to say that Maimon poses his question in order to criticize the negative conclusions of the system.

The above quotation could also be interpreted differently, as suggesting that any response to Maimon's question would require an intellectual intuition *(Intelektualle Anshauung)*, which is not the one we possess. Such an intellect comprises an intuition with a special affinity to the thing-in-itself. But such a view of Kant's response sees him as criticizing Maimon from within the Kantian system itself, for Maimon, even from Kant's point of view, presupposes in his resolution of the quid juris question that human cognition plays a part in the intellectual intuition. Rejecting Maimon's conclusion on account of its contradictoriness would amount to begging the question.

Kant's reply to Maimon's question is not satisfactory in any respect. Kant's explanation of synthetic a priori knowledge does not prove the intelligibility of the connection but simply assumes it as a fact. Indeed, this fact is *assumed* in order to explain the existence of synthetic a priori knowledge, but it is not explained. In a monistic position the quid juris question receives an answer, which proves that it is meaningful. Finally, the attempt to reduce Maimon's question to a position that requires the ability to perceive things-in-themselves, or even to one requiring the existence of an observing intellect does not tell against Maimon. For he does not accept Kant's premises with regard to things-in-themselves, a completely different philosophical position on what to him is the correct answer to the quid juris.

The *Quid Juris:* Inside Mathematics

Maimon has pointed out a lacuna in the Kantian system. But there are two ways of treating it. One is to accept Kant's answer, that is, that the system on the whole is valid and only the problem of this lacuna remains, which has to be solved somehow. Alternatively, the lacuna may be seen as a symptom that raises the need for further amendments, which already will not leave the system intact. At times it seems Maimon accepts the first option and demands that the Kantian system be repaired by adopting the assumption that both species of consciousness, the sensual and the intellectual, have the same origin and that this dualism be rescinded. On this reading, the idea of the infinite understanding, and the idea of intellectual differentials that lie at the base of the sensual phenomena, could be seen as showing Kant a way to answer the quid juris and thus a way to complete his system.

But a more careful reading reveals a different state of affairs: Maimon is not posing a question to Kant from within the Kantian system but from the perspective of an alternative system he has created. The first reading admits the ideality of space and time and also the criticism of the possibility of metaphysics as knowledge of the world of things-in-themselves separating God's world from the world of man. In the second reading, however, the quid juris challenges these conclusions.

Kant was apparently willing to accept Maimon's question, according to the first reading, despite the solution suggested by Maimon in his letter. A clear indication to this is found in the third *Critique* and also in the first *Critique* (B-29). But the second reading could lead Kant to respond by arguing that irrespective of whether he has an explanation for the relation between concepts and a priori intuitions, he cannot understand how

Maimon could possibly wish to return to Leibniz's rationalist monism, which cannot account for synthetic a priori judgments.

And if indeed the connection between concept and intuition as it appears in the idea of construction is a legitimate one, why reject the linking of the category and time? At the present junction, we have Kant's argument concerning rationalism and empiricism and we have Maimon's query about Kant's reply to the quid juris question. The challenge facing Maimon is clear: he has to solve the question of synthetic a priori knowledge without falling into Kantian dualism between concept and intuition, and without restricting geometry to an investigation of the general mode of sensation, thereby limiting the objectivity of our cognition.

This is the origin of Maimon's interest in the question of the possibility of mathematical cognition. For Maimon, in his attempt to rehabilitate metaphysics after Kant, the shift to mathematics is far from coincidental. Kant, in his "Aesthetics," begins by restricting the scope of the understanding. Geometry does not inquire into the properties of things-in-themselves but rather into the form of the sensation. It is therefore natural for a rationalist like Maimon to reassess the conclusions of Kant's analysis of geometry.

The following passage, which shall be at the pivot of this chapter, clearly demonstrates the primacy of this question as regards mathematics. Kant, as Maimon explains, asked the quid juris question about the relation between the conceptual forms and a posteriori objects. But, "as far as my opinion goes," Maimon adds: "I present, as further grounds, a fact —but not a fact about a posteriori objects (which I doubt), but a fact regarding a priori objects (of pure mathematics). Here we connect forms (relations) with intuitions, and as this fact cannot be doubted, and as it refers to a priori objects, it is certainly both possible and real. But my question is as follows: how is this to be explained? . . . Kant shows us only the possibility of his own fact, which he simply assumes. My fact, on the other hand, is both certain and possible. All I am asking is what premises do I have to accept in order to account for it?" (*Versuch* 363).

The quid juris question, we learn from this quotation, does not apply to anything still tinged with any of the doubt involved in the quid facti question. While Maimon challenges the objective validity of causal statements, he points his question at mathematics. This is a further consideration in favor of the primacy of the question as regards mathematics. Thus, although Maimon adopts Kantian terminology in formulating the difficulties involved in the relation between categories and the a posteriori, his discussion of Kant leads him to pose a new question or, rather, to transpose the question from one field into another. Maimon generalizes the Kantian question by a sharp formulation referring to the concepts and facts of mathematics. This can give us some idea of the damage resulting from the typical neglect in scholarship of Maimon's latching onto mathematics, especially regarding his mathematical insight.

Pure Cognition and a Priori Cognition

Maimon was not a mathematician, but he was familiar with contemporary disputes on the subject of calculus and its relation to geometry, which was then considered the paradigm of mathematics. He had a thorough understanding of the different mathematical trends that had existed even before they surfaced in nineteenth-century mathematics. Mathematics serves as Maimon's starting point, and his grasp of mathematics places him on a par with the leading mathematicians of his generation. He innovatively advanced the possibility of mathematically fertile non-Euclidean geometries well before Gauss. Mathematics, for him, was not a tool to be used in the physical sciences, but an independent, autonomous inquiry, with a status independent of its application to the natural sciences. This position must be appreciated against the historical background of Maimon's time, which saw the beginnings of the algebraization of mathematics and the beginning of its detachment from geometric intuition and mechanics. In Maimon's time, negative numbers became a standard part of mathematics and mathematicians were forced to accept complex numbers, although still regarding them as artificial creations. Maimon was one of the first to realize the conceptual implications of the recognition of these mathematical entities. His complete commitment to metaphysical questions, however, enabled him to apply these implications to the thorough examination of the Kantian system.

Maimon's starting point is the classification of all mathematical knowledge into different species. In the *Versuch,* Maimon distinguishes between two types of a priori cognition, which already figure in his original formulation of the quid juris question:

> The pure is that which is a creation of the understanding only (and not of sensibility). Everything that is pure is a priori, but the reverse is not true . . . the circle is an a priori concept, but this does not mean that it is a pure concept, because the intuition (which I did not create from within myself, according to a rule, but rather it came to me from a different, though a priori, source), must serve as the foundation of this concept. And pure concepts, according to my explanation (those that do not contain any intuition, not even a priori intuition), can only be correlative concepts, as the concept is nothing but unity in multiplicity, but the multiplicity can only be considered unity if it is necessary to regard its parts as mutually dependent, or, at least, unilaterally dependent. (*Versuch* 36, 56)

The a priori, then, can be divided into the pure and the sensuous. The first indicates an aspect of our cognition, which does not refer to sensibility in any way: neither to specific sensations nor to the imagination, nor even to that which constitutes a condition for the sensation of objects—namely, space and time. This level of the a priori is completely conceptual, which is why Maimon calls it pure. The second type is the a priori, which involves space and time and refers to the forms of sensation.

This distinction is reminiscent of Kant's distinction between thought *(Denken)* and cognition *(Erkennen).* Thought does not involve any reference to sensation or to the conditions of sensation, which is why we can think of the thing-in-itself. In the case of cognition, though, we demand that our cognition refer to a possible object of experience. In Kant's opinion, we can think of anything that does not involve a contradiction, but in order to be aware of something, we have to refer to the conditions of sensation. However, as we shall see, the range and philosophical significance of Maimon's two types of cognition are completely different from the types discussed by Kant.

In his *Logik,* Maimon expands and develops this distinction. Pure cognition and a priori cognition are two distinct levels of cognition, to which he adds two further levels. These do not entail four mathematical theories but rather four different types of mathematical inquiry.

The General Theory of Magnitudes
(Die allgemeine grössenlehre)

Maimon describes this level as follows:

> The general theory of magnitudes . . . refers only to all the kinds of forms (Formen) or relations (Verhaltnisse) in which magnitudes can be thought of at all, regardless as to whether or not these forms, if applied to specific objects, will ultimately lead to indeterminable (infinite) or even to impossible magnitudes. (*Logik* 1)[1]

Or:

> At this level, we abstract not only from all the empirical objects to which it applies, but even from all the objects determined a priori by pure mathematics. (*Versuch* 56)

The first level is a general theory of magnitudes. Maimon emphasizes the point that this level analyzes possible relations between magnitudes. This level is not restricted to any particular magnitudes, and Maimon is even prepared to extend it to impossible magnitudes—namely, as he notes in several places, to complex numbers. This definition clearly shows Maimon to be party to the endeavor to interpret algebraic truth, a project that gathered momentum later on in the work of Peacock and Hamilton and that led to the concept of algebraic structure. However, in order to appreciate the uniqueness of Maimon's position, one must examine the notions of relation *(Verhaeltnisse).*

On the basis of the account Maimon provides in the appendix, relations should be understood as standing in contrast to inner properties *(innere Merkmale).* In making relational judgments, we determine an object in relation to another object. There are two ways of doing this: either the comparison that supplies the relation is drawn on the basis of the object's inner properties or it is not based on given inner features but rather determines the objects by means of the relation. In the first case, as opposed to the second, the relation or the comparison is reducible to single place predication. For instance, when we assert that the table and chair in front of us are of the same color, we can render this statement as a disjunction of the type: the chair is black *and* the table is black *or* the chair is blue *and* the table is blue, etc. In the second case, the relation is irreducible in this fashion, since it is original and independent of the object's characteristics; on the contrary, the statement asserting a relation between numbers is constitutive of the numbers specified.

Mathematics at this level deals with relations in the second sense, and arithmetic is the paradigm case. Natural numbers are not objects that exist independently of the relations between themselves, but, on the contrary, the relations between them actually constitute them: *"The numbers one and two in and of themselves (abstracted from their application) are objects of pure arithmetic which are* a priori *determined by relations. They have no inner properties apart from these respective relations"* (*Logik* 1 n).

This idea of Maimon's implies a far-reaching innovation. Metaphysics before Maimon regarded the category of relation as a category of a low degree of reality. Following Aristotle, relational statements were regarded as derivative statements, which had nothing to do with the essence of things. Maimon, with his idea of correlative statements, as reflected in the first level of inquiry, completely reverses this conception. Relation can act as the foundation of an object and, as such, is real and primary.[2]

Maimon draws a distinction between logic and the general theory of magnitudes (*Logik* 2). Logic deals with all objects as such, with no reference to their determination; in other words, logical statement forms are true no matter how they are instantiated. The laws reigning in logic are the law of contradiction and the law of identity.[3] Thus, according to Maimon, logic cannot be creative, for the laws of logic alone cannot constitute objects. In order to constitute objects we would have to be able to distinguish them from each other, and we cannot do this so long as we confine ourselves to the laws of logic alone, such as the laws of contradiction and identity or even all of Aristotelian logic. Logic is a restrictive science, which determines what we are not allowed to claim (see the introduction to *Givat Hamoreh*). It is thus devoid of creativity.

In light of the distinction between logic and mathematics, and given later developments in the field of algebra, one can attempt a more precise definition of this level. For Maimon, pure cognition has *two* distinct aspects: one is the abstract manner of dealing with the relations between magnitudes in general, and the other is the determination of objects by means of these relations. In the first sense, the objects must be given irrespectively of their relations, and in the second sense, the relations constitute the objects. The first case is, in a manner of speaking, the laws of magnitudes, whereas the second case refers to the objects created according to these laws.

The Theory of Pure Magnitudes (Die reine grössenlehre)

The objects of inquiry at this level are those objects determined by impure a priori construction. This level refers to objects only in so far as they can be constructed. Geometry, as a science that constructs and realizes its objects in a priori intuition, belongs to this level. Maimon considers this level to be subordinate to the general theory of magnitudes. In fact, it is obtained by restricting the general theory of magnitudes to everything that can be constructed a priori.

General Mathematics (Die allgemeine Mathematic)

This level is similar to what we call applied mathematics. At this level, we bring pure mathematics to bear on empirical objects. Maimon therefore adds the following important comment on the nature of this level: "It may employ pure mathematics only assuming that the conditions which apply to the empirical objects do indeed correspond to or coincide with the conditions which apply to the pure objects" (*Logik* 2).

Mathematics applies to empirical phenomena only if one supplements premises as to the specific laws governing these phenomena. If the laws of nature were any different, then the contents of this level would be different. Therefore, Maimon explains, all inquiries at this level are hypothetical. As an example, he presents the fact that we may deduce the statement that the pendulum swings in a cycloid only by presupposing the particular form of Newtonian gravity (*Logik* 2). The modern counterpart of doing mathematics at this level is applied mathematics, comprising mathematical physics.

Practical Mathematics (Die praktische mathematic)

In this level, involving a higher degree of reality, we undertake to mechanically construct objects, which obey a predetermined condition (*Logik* 2). In fulfilling this condition, we make certain physical assumptions. This level can be compared to constructing bodies defined by geometry, or even to engineering and architecture.[4]

The main point of the quid juris question, then, is the distinction between the first and the second levels of studying mathematics and the relation between them.

The Question

It is the distinction drawn between types of mathematical cognition that enables us to formulate Maimon's quid juris question. Thus, the question regarding the relation between the categories of the understanding and the forms of sensibility is generalized into a demand to understand the connection between pure a priori cognition and a priori cognition, which draws on intuition. More explicitly, the question concerning the connection between causality and time is transposed into a question about the relation between natural numbers and space and time. We acknowledge the connection between pure cognition and a priori cognition, but how do we make such a connection intelligible to ourselves?

This way of presenting the issue needs sharpening. It should first be noted that not every connection between one form of cognition and another raises the quid juris question: when we apply logical rules of inference in arithmetic, or on the axioms of geometry, we do not encounter any particular problem. Logic, for Maimon, comprises all statement forms that are true under any instantiation. So however we instantiate the formula "*X* is *X*," we will not run up against a troublesome synthesis. For it is a necessary condition for anything to be an object that it abide by the rules of logic, and the instantiation merely confirms this.

A somewhat more interesting case is the relation between logic and intuition when the latter is intended to aid us in deciding as to the validity of an argument. When we fit a Venn diagram to an Aristotelian syllogism and determine the validity of the argument form by inspecting the diagram, an intuition that is not empirical and is not concerned with the color of the circles or even their size, we are connecting logic with intuition. However this connection does not raise the quid juris question. What we see in the diagrams reflects the situation in respect to the validity of the syllogistic form tested, but the validity of the forms is not essentially based on the diagrams.[5] On the other hand, the synthesis between space and arithmetic or between time and causality is much more intensive. A logical inference form does not require intuition for its very validity, but when we come to the connection between construction and geometrical truth, Kant insists that without this connection, this truth would not be possible. The quid juris question is posed concerning the last and not the first example.

The instructive example Maimon gives of a bond between pure cognition and an a priori cognition is the proposition that the straight line is the shortest line. In this proposition there occurs a synthesis between two constituents. On the one hand, there is the straight line, which, as far as Maimon is concerned, is defined as a line that appears in intuition as having a fixed direction. This is the straight line encountered in geometry, hence it pertains to impure a priori cognition. One cannot ask for an *authorization*—on what basis do we apply the concept of an *identical direction* to the relevant intuition? The drawing of the line is already governed from the start by the rule of the identity of direction.[6] On the other hand, the property of being the shortest is an attribute or a relation pertaining to the magnitude of the straight line, and therefore it belongs to pure cognition. Propositions of pure cognition do not raise a quid juris problem, for the connections they affirm are necessary for the constitution of the entities referred to. But this is not the case with regard to the proposition in question. In Maimon's words: "Here the inner (the thing as it is in itself) precedes the outer (the relation to other things); on the contrary, that is to say, without the thought of a relation there is no concept of magnitude (in pure arithmetic), whereas in geometry we are presented with objects, i.e., forms that are determined by their positions *before* they are brought under the category of magnitude" (*Versuch* 69; emphasis added).

To better understand this point, we have to recall what Maimon says on the first level of inquiry, the general theory of magnitudes. The proposition "4 >3," which lies at that level, is not problematic, because here the objects and the relations come into being inseparably: "The understanding, in my opinion, is merely the capacity to think, that is, the capacity to create pure concepts by making judgments. No objects are granted to it as material on which the mind has to operate . . . it is wrong to think that it is necessary for things (real objects) to be prior to the relations between them. The concepts of numbers are only relations, which do not presuppose any real objects, as these relations are the objects themselves" (*Versuch* 190).

The propositions "3 >2," "12 >5," and so on, which we encounter in arithmetic, are true not by virtue of any intrinsic property of 2 and 3. On the contrary, 2 and 3 are both mutually defined by the fact of their standing in this particular relation to each other. The relation is not an external supplement to numbers that exist as objects independently of it. But the relation between intuitions and concepts that figure in geometrical judgments, and which Kant's theory requires that it bring together a conceptual term and a nonconceptual term, raises a difficulty.

The difference hangs on the fact that the intuitions, even when a priori, are given to us independently of the concepts of magnitude that we apply to them. From the way the quid juris question is presented we can tell that it is not merely a question about the relation between a totally unconceptualized given and the concept. The question does not imply that the geometrical straight line is a pure intuition devoid of all conceptualization. As mentioned, the straightness in this example is given through the identity of direction. But even if we do concede that there is some conceptualization in every intuition (for example, that through two points only one line can be drawn), then, nevertheless, we are *binding* to it an *additional* characteristic. How do we account for the possibility of such a bond, that is, the possibility of applying a pure concept to intuition, which is a priori but not pure?

The commentary on Maimon has completely ignored the formulation of his question in terms of the relation between pure cognition and a priori cognition, and this despite Maimon's repeated emphasis of the difference between these two modes of cognition. Instead, scholars have tended to conflate his question with the Kantian question regarding the possibility of synthetic a priori knowledge.[7]

There is no denying that Maimon's wording furnished ample footing for such a construal. The following excerpt illustrates the great latitude Maimon permits himself in presenting his opinions: "Thus if we assume that time and space are a priori intuitions, then they are merely intuitions and not a priori concepts: they present us simply with the elements of the relationship and, by means of these, the relationship itself, but they do not present the truth or the justification for employing this relationship. The following question thus arises: how are synthetic judgments possible in mathematics, or, how is it that we acknowledge their existence?" (*Versuch* 60).

Here there is a strong suggestion that Maimon is equating the initial question—which is the version I am adopting here—with the question at the end of the passage, which is about synthetic a priori judgments. One should not, however, be deceived by the similarity in the phrasing of the two questions. One should just give full attention to the wording: "they present us only with the terms of the relation and through them with the relation itself, but they fail to expound the truth and the justification in putting this relation to use."

It is evident that the terms of this relation differ from numbers; precisely in that numbers cannot be thought inseparably from their relations. Here, suggests Maimon, we are invited to use the relation, given the terms, but we are not shown the *authorization* for such use. How are such judgments justified, or shown to be true, asks Maimon. And his question, his quid juris, is patently not Kant's question but rather a query concerning Kant's *solution* to the synthetic a priori problem, the solution requiring linking pure concepts to a priori intuition.

To my knowledge no one has linked the dualism in mathematical cognition with classical dualisms such as the link between causality and time and the mind-body dualism as Maimon did. And in general the harmony between different branches of mathematics is not recognized as problematic. Of course there were allusions to this problem before Maimon and after him. Spinoza's parable of the connection between the equation of the circle and a drawing of a circle illustrates that two apparently alien expressions can actually be expressions of the same thing. This parable that was intended to resolve the thought and extension dualism and the mind-body problem is pertinent to Maimon. Maimon also draws an analogy between mathematics and the mind-body problem, and it is possible that Spinoza's solution, which utilized the concept *expression*—for example, in affirming that the attributes extension and thought are each complete expressions of God—is what prompted Kant to call Maimon's system Spinozism.

For Spinoza, the connection between a circle and its formula is an analogy, whereby we are meant to understand how two things can ultimately be expressions of the same thing. Now, in order to work, this analogy presupposes that this very connection between the circle and its formula is not problematic. But for Maimon it is precisely this connection that is troublesome: how can there be a connection between two things that are conceptually totally foreign to each other?

A remote source of Maimon's question can be found in Aristotle, who thought one could not prove a geometrical theorem arithmetically, and vice versa, because the propositions are of diverse kinds. Even as one cannot learn from camels about ants (and conversely), it is impossible to infer from shape about number (and vice versa). Aristotle discredits the fact on the ground that it cannot be rendered intelligible. His position is supported by his distinction between knowledge of the fact and knowledge of the reasoned fact, thus corresponding to the difference between the quid facti and the quid juris. Maimon, in his critique of Kant, can therefore be seen as translating Aristotle's insight into the transcendental idealistic idiom. Maimon is aware of the *fact* of the connection between geometry and arithmetic, as Aristotle could not be. But Aristotle's perplexity remains: how can there be such a connection?

Maimon was deeply immersed in Aristotelian thought, as can be seen from his work on Aristotle's categories. The air was still saturated with Aristotle, as is attested by Leibniz's work and influence, and even more so by Wolff's work. The latter embraced the distinction between knowing a fact and knowing it by understanding and defined philosophy as the knowledge of the reasoned fact. Clearly, for Maimon's generation the quid juris goes to the very essence of philosophy.[8]

Maimon's question is directly connected to the question of the applicability of mathematics, but this is not a question about the relation of the Platonic world to the temporal world, since Maimon, as a radical idealist, doesn't accept the separate existence of such a Platonic world. The question is not aimed at the temporal world at all but at the relation of pure mathematics that doesn't require the visualization of space and time to the mathematics that does. Therefore, Maimon's question is not about the application of mathematics in the natural sciences but about the *application of mathematics in mathematics.* This projection of the question has a very respectable heredity. Plato's parable of the divided line moves the relation between the phenomena and the world of thought into the world of thought, since the relation between the two, the world that is seen *(horomena)* and the world of thought *(nooumena),* is conceived as identical to the relation between dialectics *(dialektike)* and the sciences *(mathemata).*

This point can be accentuated from a different angle. A particularly interesting mathematical phenomenon is the amazing harmony between the various branches of mathematics. The harmony between number and plane expressed in the arithmetization of the plane by Descartes and the applicability of analysis to number theory is something of a mystery.

For mathematicians, this connection is so obvious that they do not usually stop to think about it. Let's take, for example, the connection between planes and complex numbers. We define the operations of multiplication and division for real numbers—which today are granted with no reference to geometrical straight lines—followed by the operations of squaring and deriving square roots. We now come to the meaningless expression: "the square root of minus 1." We can overcome this problem by means of algebra, at which point we find that we have reached a plane. The line and the real number lead us to a plane.

Furthermore, a brief search reveals an even deeper connection. It turns out that multiplying by the root of minus 1 is none other than the intuitive operation of revolving a point at an angle of 90 degrees. We are familiar with such an operation irrespective of the operation of multiplying by the complex numbers, but, surprisingly, this multiplication has geometrical consequences. We make a movement in one medium and discover that this movement corresponds to an operation in a different medium. We know there is no conceptual connection between numbers and planes, yet they are nonetheless connected in an unusual way.[9]

Maimon's motivation for dealing with the quid juris question is of a metaphysical nature. In translating this question into the language of the *Critique,* he arrived at the issue of schematism and the question about the relation between the categories and time. But his dissatisfaction with Kant's solution and his endeavors to question the coherence of the Kantian system led him to the question of the relationship between pure concepts and the impure a priori. After refining this question, Maimon was led to formulate a philosophical-metaphysical question within mathematics. The classical question of the relation between God and the world becomes the question of the relation between numbers and space and time.

3

Maimon's Ladder

Maimon's solution to the quid juris question appears in the following passage, which is often referred to in the literature on Maimon: "According to the Kantian system, which states that the sensibility and the understanding are two sources of our cognition, that are completely different from each other, the question, as I have demonstrated, is insoluble; but according to the Leibniz-Wolff system, both of these elements stem from the same source of cognition (the difference between them being only a difference in the degree of the completeness of this cognition), and the question is easily resolved" (*Versuch* 64). As long as we regard sensibility as independent of the understanding, we cannot comprehend the possibility of applying concepts to sensual intuition. The connections between them can only be explained by assuming that they both derive from the same source.

This is the line usually taken by commentators on Maimon's solution to the quid juris question. However, such a formulation is too schematic and metaphorical, and the central notion it employs—that "both of these elements stem from the same source"—is vague. Other metaphors found in Maimon's writings and in the commentary on him

—such as "the difference between an intuition and a concept is one of degree rather than kind," or "intuition is an imperfect understanding"—do not improve matters. All these formulations are not clarified even by appeal to Maimon's theory of differentials, which, according to this doctrine, constitute the intellectual element whose integral is intuition. Similarly, the idea that for an infinite understanding there is no form-matter dualism is also unhelpful and demands clarification like the other metaphors. Kant's system amounts to an intentional abandonment of Leibniz's dogmatic rationalism, "which made all phenomena conceptual" (*Critique* A-270, B-326). If there is indeed anything innovative and original in Maimon's position, it cannot be a mere reiteration of Leibniz's ideas.

Maimon's metaphorical phrasings should be construed as mere headlines, as he himself suggests. In order to explain Maimon's response, we have to try to extract as clear a picture as possible from the depictions and examples he discusses. This is no easy task, if only because already at first glance Maimon's solution to the quid juris question strikes one as containing an abundance of ideas, apparently suggestive of more than one response.

Maimon and Formalism

Formalism is a school of thought that often comes up in the course of the critique of Kant's concept of a priori intuition. Russell believed that Kant's view that the axioms of arithmetic are synthetic was a result of Kant's possessing a rather rudimentary logic. But he also added that the use of construction in geometry was merely a result of the fact that Euclid's axiomatization was incomplete, and so he occasionally resorted to drawings in order to complete proofs. In Russell's opinion, Hilbert's axiomatization of geometry, along with Frege's work, completely canceled the need for mathematical intuition. Cassirer took a similar approach in interpreting Hilbert as having demonstrated that geometry is the study of relations: "Intuition seems to grasp the content as an isolated self-contained existence; but as soon as we go on to characterize this existence in judgment, it resolves into a web of related structures, which reciprocally support each other" (Cassirer 1923, 94). Bergman, too, on the basis of passages quoted from Maimon, equates Maimon's views with this type of formalism.

Hilbert's work on the foundations of geometry had a decisive influence on many mathematicians' conceptions of geometry. Until the publication of this work, geometry was perceived as the study of the consequences of certain axioms, whose clarity and evidence impose recognition of their truth on us. With this view, we start with simple axioms, recognized as true by the light of reason (or by synthetic a priori intuition) from which we infer, in accordance with the laws of logic, one theorem after another. At the same time, we assume certain concepts as basic and use them to define the rest of the concepts involved. Formalism proposes a different approach to geometry. From this view, the sharp distinction between definitions and axioms becomes blurred. Thus, unlike the traditional view, axioms do not describe truths about concepts whose prior meaning is hinted at; rather, they strictly define these concepts by determining the connections between them.

A definition in this sense is not an explicit definition in which the concept being defined is presented as equivalent to the meaning of the defining phrase. For this reason, Hilbert's definitions are usually referred to as implicit definitions. This title refers to the kind of definition common in mathematics, where what is being defined has no status independent of its definition. The classical examples of such definitions are those of functions by means of the equations they satisfy. But here we find an expansion of the idea of the implicit definition: when we claim that "between any two different points there is a third point," we define the relation *between* and the concept *point*.

One notion central to this approach, which is a corollary of the new notion of definition, is that mathematical operations can ignore the question of the system's application in the world, or even the question of its correspondence to our intuition. When we study a system of axioms, all we need to do is determine whether or not it is formally consistent. This leads to a number of ideas, usually associated with the relation between formalism and the concept of truth:

1. Consistency is a sufficient condition for truth, or at least for mathematical truth.

2. The entire purpose of mathematical inquiry is to establish relations of entailment between sentences: if certain axioms are true, then other sentences are also true. Euclidean geometry thus comprises conditionals, which are unconditionally true.
3. Mathematics is the study of meaningless symbols. All that is required of a game of meaningless symbols is that it be interesting. Consistency is an essential requirement because, without it, any sentence could be proved, in which case the game would not be interesting.

I do not intend to enter into the history of formalism, much less in its Hilbertian version, which has undergone diverse developments.[1] The points above are important because one can find clear suggestions of all of them in Maimon's work. An approximation to *A* and *B* can be found in the very possibility of non-Euclidean geometry, which Maimon explicitly discussed. Kant had already noted this possibility in arguing that geometry is not analytic and that therefore negating it would not involve us in a contradiction, but Kant never imagined the possibility of a fruitful non-Euclidean geometry. In fact, he was convinced that the parallel axiom, surrounding which non-Euclidean geometries were discovered, could be proved, and he even attempted to derive it from Euclidean principles.[2]

Maimon's prediction of the possibility of non-Euclidean geometries, in contrast, is not an incidental consequence of his system, or simply a lucky guess, but is part of a wider picture: "Logical truth is the connecting of the objects of thought (concepts) in accordance with the laws of the understanding. The axioms (insofar as they are not the product of any linking) are the elements of truth but not the truth itself. The results that stem from the linking are the products of truth but not the truth itself for, in accordance with my explanation, truth means only the course of reason or its legitimate manner of thinking and not the principle from which the understanding proceeds nor the result it finally attains" (*Versuch* 148). On this view, truth is an act or a move of the understanding.[3] What is crucial is the derivation and not its antecedent and consequent. These are only important when we come to practical implementation. But they also have "a use within thought," distinct from practical uses. Maimon explains: "If Euclid had assumed false axioms instead of his metaphysically true ones, I am certain that he would have left the world a work that is in no way poorer than or inferior to the one that has come down to us" (148).

Maimon even takes a step toward realizing such geometry: "I assume, for example, that the outer angle of a triangle is not equal to the sum of the two opposite inner angles but to this sum plus one half; it necessarily follows that the angle at the center of a circle is not twice as large (as it actually is) but three times as large as the angle on the periphery of the circle, and so forth" (148). It is obvious from this passage that Maimon noticed the possibility of developing the conclusions of non-Euclidean axioms, but instead of applying this idea to the parallel-postulate, he started from the theorem of the sum of the angles in a triangle.[4] Maimon was aware he was on the brink of discovering a different mathematics. This activity, so he explains, "furthers the intellect, by offering us new horizons in which to employ it," and he adds, parenthetically, "we could, by way of example, invent a new mathematics" (150).

It thus appears that Maimon also believed consistency was sufficient to establish a new science of mathematics. However, in spite of the affinity of the first principle mentioned above in the discussion of formalism, with the idea that consistency is a sufficient condition for mathematical truth, or, in his words, that "the meaning of truth is only the course of reason," one should not regard formalism as the way in which Maimon hoped to free himself from intuition. The idea of different geometries does indicate that Maimon was of the opinion that geometrical propositions are conditionals that can be proved by means of logic, but it is a far cry from this idea to Maimon's position discussed below, according to which intuition is an incomplete concept.

At this stage, we can make some observations of the closeness between the two positions. To the extent that we employ intuition in the course of proofs, Maimon's argument that intuition is incomplete understanding can be identified with his view of truth as an intellectual process. Accordingly, the intuition that we refer to in the course of a proof is, in fact, an unconscious conceptual operation—a sort of blurred image of a proof or of "an intellectual process." But Maimon also draws a connection between the intellectual and the intuitional at the level of axioms—that is, where there are no proofs. In order to explain this aspect of his thought, one cannot resort to its connection with formalism. Maimon recognized the possibility of salvaging the pure activity constituting demonstrations of mathematical theorems by regarding them as conditional propositions whose antecedents are axioms, or proved theorems, and whose consequents are provable theorems in the system. But he insisted that such a salvaging should not explain their use, or their "metaphysical truth" (148–49). Even if geometry *can* be dealt with formalistically, one does not thereby capture the essence of geometry.

Treating the propositions of mathematics as conditionals indeed makes it possible to ignore the axioms, because one is focusing on the entailment only; but this is far from a response to the question of the relation between a priori cognition and pure cognition. For in the same way that we liberated ourselves from intuition here, one could turn physics into a deductive system and be exempt from experimenting. Deductivism trivializes epistemology, and even as one cannot consider it a mode of rationalism, one also cannot construe the transposing of geometry into conditional propositions as a claim to a single source for both intuition and concept.

Another issue pertaining to Maimon's relation to formalism concerns what Maimon called symbolic cognition. For Maimon, this is an additional source of cognition alongside the pure and a priori modes of cognition. In the course of the *Versuch,* this source of knowledge assumes an ever-growing importance, so that ultimately Maimon devotes a special lengthy appendix to it in the *Versuch,* under the title "Über symbolische Erkenntnis." This third mode of cognition, which approaches what I referred to above as the third characteristic of formalism, merits special examination. I shall confine myself, however, to asking to what extent this third cognition can be identified with pure cognition.

The objects of symbolic cognition are the symbols themselves and the relations between them (see *Versuch* 265). Symbolic cognition does not refer to sensibility (beyond the symbols themselves), and thus it is not a cognition of the second, a priori level. But Maimon does not only *add* symbolic cognition as a type of cognition distinct from pure cognition and a priori cognition; he also explicitly states that consistency—which should be sufficient, were mathematics the mere study of symbols—is not a sufficient condition for truth (*Versuch* 101).

The relationship between pure cognition and symbolic cognition is illustrated in its full force in Maimon's coin analogy. The essence of the coin comprises two components: first, every coin has a conventional component, reflected in the coin's designated value, which is granted by the institution determining that a particular object is to be regarded as a coin. In addition, every coin has a value equivalent to the value of the material from which it is made. Maimon tries to capture certain elements of the relationship between the nonsensible and the sensible by means of a comparison with the two components that constitute the coin. He explains that symbolic cognition is helpful in discovering the truth; when we employ a convenient form of formalism and do not allow ourselves to be led by intuition, we can expose conceptual relations that we could not have noticed by means of intuition. Maimon notes as examples algebra and the advantages of analysis over geometry (see, for example, *Versuch* 412). However, alongside this alleged advantage of symbolic cognition one should not overlook its basic limitations: "Thus, without much toil, mechanically, as it were, one can discover the truths, be they even most concealed. However, a new difficulty thereby emerges, namely, sometimes one reaches symbolic combinations or formulae that are meaningless numbers, tangents, the cosine of a right angle, etc., in mathematics. Symbolic cognition is indeed an excellent means for inventing truth, but we must be extremely cautious in employing it. At every step, we must ask ourselves, as the statesman would put it: can this symbolic coin be cashed? If not, we will find ourselves faced with the most peculiar ideas, from which we will not be able to extricate ourselves later" (412).

According to Maimon, a symbol with no corresponding intuition is problematic and devoid of content. From this point of view, the symbol is different from the coin; the coin's very nature lies in the disparity between its symbolic value and its value as merchandise, which is the value of the material from which it is made. When this disparity is at its maximum, the coin is truly a coin in the full sense of the word. In contrast, when truth is entirely symbolic and has nothing to do with intuition, then "we reach symbolic combinations or formulae that are meaningless."

The relation between symbolic cognition and a priori cognition can help us distinguish between symbolic cognition and pure cognition. Maimon clearly states at the end of his analogy: "In this case, thought is actually distinguished from the coin, in the following sense: as, in the coin, form in itself is abstracted from substance, it has no value whatsoever, and can thus only be used as a symbol in a manner agreed upon in the land, but it cannot be put to general use, whereas the matter retains its general value. In the case of truth, however, the converse is true: the form has a general value, to the extent that by means of it a real thought is always created, whereas the matter is of value only for us, and not for every thinking being" (415).

As stated, intuition controls symbols and grants them their value, and when the symbolic goes beyond the intuitional, it becomes fiction. However, the above analogy reveals that when *thought* goes beyond intuition, its objectivity is not affected. These two points conjointly produce the following hierarchy as regards the value of different types of cognition: from the point of view of Maimon's rationalism, pure cognition is the most valuable type. It "ensures the value of every thinking being" and is thus responsible for the objectivity of our cognition. One step beneath this type of cognition lies intuitive cognition, which is of value for us only ("the matter" in the above quotation). And at the lowest level we find symbolic cognition, which is controlled by intuition.

In addition to the above points, there are methodological considerations against the classification of Maimon as a formalist. Hilbert's formalism, as interpreted by Russell and Cassirer, dispenses with sensibility, which transcends the symbolic. Yet what is left over once one has abandoned sensibility is not of much avail to the rationalist. The obvious candidate for substituting the mind is a meaningless system of relationships, which the rationalist is not warranted in identifying as the intellectual "something" that lies at the basis of everything. Maimon's view, according to which everything intuitional is an expression of the intellectual, seeks to grant both the intuitional and the intellectual a greater significance. In the context of Maimon's thought, the intuitive takes on an objective significance by being a picture of the objective; and the intellectual assumes an ontological dimension by being the foundation of the intuitive.

Maimon's extreme idealism may further our understanding of this ontological element. The formalist in mathematics is usually a realist in physics; if he were not, all he would be left with is nothing but symbols and intuitions, which would bring him too close to psychologism. If the materials Maimon was working with were only signs and intuitions, he would not possess sufficient tools to challenge the subjectivity Kant forces on us by means of his thesis on the ideality of space. The Maimonic world is furnished with signs, intuitions, and thought. As we shall see in the following chapter, the objects' role of guaranteeing the objectivity of our cognition is fulfilled by thought. This is why the distinction between signs and thoughts is essential as far as Maimon is concerned.

A third issue that links Maimon's position to formalism has to do with his claim that the concepts in propositions within the space of pure cognition define each other. This may lead one to assume that Maimon's view is not unlike Hilbert's manner of perceiving the relation between axioms and definitions. But such an analogy is problematic. It involves certain difficulties regarding the distinction between arithmetical propositions, which are paradigmatic cases of correlative propositions, and geometrical axioms, as perceived by Hilbert. As far as the latter are concerned, the formalistic account is somewhat vague. On the one hand, the formalist argues that the concepts of a straight line and a point are implicitly defined by the axioms of geometry, thereby providing us with definitions of a straight line and a point. On the other hand, however, the formalist subscribes to the view that mathematical axioms and proofs are a formal play of symbols, which requires only the minimal condition of consistency.

But definitions, including implicit definitions, can never be empty games; the implicit definition of a function, or its definition as the requirement of satisfying a differential equation, is not a play of symbols, because we have theorems of existential import and uniqueness attached to the definition. Hilbert extended the concept of mathematical definition, but by this extension he forfeited an important characteristic of the implicit definitions found in mathematics. Any well-formed sequence of symbols can be considered as a formal system that defines its terms, but not every combination of symbols in which the function sign occurs defines a function.[5] There is no such vagueness when it comes to correlative propositions. Maimon regarded these, like analytical judgments, as true propositions in every sense of the term. Indeed, Hilbert's geometry requires an interpretation, while the propositions of arithmetic do not.[6]

Maimon anticipated formalism, which began to flourish only toward the end of the nineteenth century, but, contrary to what Cassirer and

Russell thought, he did not regard it as the solution to Kant's problem. Formalism ignores intuition and does not reveal its roots, and so long as we have not explicated and resolved this issue, we lack the answer to our question. One may draw the comparison differently: the formalist seeks to save geometry as a deductive science, without making it dependent on questions of applicability and on the repercussions of changes in the empirical sciences. This rescue attempt regards mathematical proof as the most quintessential aspect of mathematics, and this aspect is preserved in the abstracting from the content of concepts and of fastening only on the *relations* involved. Maimon, by contrast, discerns the philosophical value of mathematics *not* in the deductive system but in the process of forming concepts: "If we examine mathematical method, which is praised so highly, we will realize that it is not as valuable as people think. For by means of proof one can proceed from error to error, just as one can reach truth from truth. It is not mathematical method that is so valuable, but rather the revelation of the principle of human thought, expressed in forming mathematical concepts and their interrelations by the understanding and by reason" (*Versuch*). The formation of concepts brings us close to Maimon's answer to the quid juris question, since it is here that the intellectual essence underlying intuition is revealed.

Intuition as a Picture

The following passage embodies the essence of Maimon's response, which is completely different from formalism:

> Here we have an example of the way in which the understanding can transform a reflective concept into a rule for creating an object. For in order to create a straight line as an object, the understanding thinks in terms of the rule that this line must be the shortest distance between two points (for it cannot turn the property that it must be straight into a rule, as the property of straightness is an intuitional property and therefore lies beyond the bounds of the understanding). The property of being the shortest is really a reflective concept (a relation of difference in magnitude). One could have recognized this in advance as regards magnitudes,

> before employing them in the context of intuition, for magnitudes become objects only by means of such relations. In this case, unlike in the case of other objects, the internal (the thing-in-itself) is not prior to the external (relations to other things), but, on the contrary; in other words, without a considered relation, there is no object of magnitude (in pure arithmetic, as opposed to geometry, which presents us with permanent objects—that is, forms—before subordinating them to the category of size). The property of the straight line is, as it were, the picture of that correlative concept, and can thus not be employed in the same way as one can employ a concept in order to derive certain conclusions. If we survey all the propositions about the straight line, we will find that these propositions are derived not specifically from the straightness of the line, but from its being the shortest, in the same way that one cannot conclude, from all sensible intuitions, anything other than that they are what they are. Yet the fact that we recognize this proposition by means of intuition before even proving it, is based on the fact that we perceive in it the picture (which may be clear and not distinct) . . . The above would appear to be a paradox, for one would generally tend to suppose that the property of the straight line is, in this case, an internal property (the relations amongst the parts), whereas the property of being the shortest is an external property. However, careful examination reveals precisely the reverse to be true; that is to say, the property of being straight, or the unity of direction of the parts, already presupposes their having come into existence. Therefore the definition of the straight line will not do either. Wolff's explanation cannot save us from this difficulty, because the similarity between the parts and everything else already presupposes the lines. However, the property of being the shortest distance is founded in the very *existence* (Enstethung) and is, at the same time, an internal relation. (*Versuch* 68)

The ideas articulated in this excerpt do not appear anywhere else in the *Versuch,* and to my knowledge never recur in Maimon's late work. Here he provides a clear example of how to interpret his metaphorical descriptions, such as "the sensible is a picture of the intellectual," or "the intellectual and the sensible branch out from the same trunk."

The statement "the straight line is the shortest line" is also Kant's example of a synthetic a priori judgment, and it is about this connection that Maimon asks the quid juris question. This facilitates the comparison between Maimon and Kant and enables a close examination of Maimon's main gesture. On Kant's analysis, when we claim that a straight line is the shortest line, we are adding a further property to that of its being straight. Maimon understands this example differently. According to him, the intuition in question is not supplemented to the reflective concept in order to form a synthetic judgment; rather, "it is a picture of that correlative concept," that is, it represents the concept on which it is founded. What we see as a straight line—that is, a line with a single, fixed direction—is none other than *being the shortest route.* In other words, Kant reads the proposition "the straight line is the shortest line" as a proposition in which "being the shortest" is the predicate of the subject "the straight line." In this respect, it is no different from the proposition "all men are mortal." Maimon ultimately reads the proposition as one of identity: being a straight line *is* being the shortest line.

We must now try to understand what is meant by this identity, for an understanding that is consistent with Maimon's principles might prove to be the key to his entire system. To state that the straight line is the shortest line amounts to claiming that the property of "straightness" is reducible to being "the shortest." Indeed, if I had chosen to construe this identity in analogy with the identity between 2 + 1 and 3 (or that between the evening star and the morning star), I would have missed—owing to the symmetry between the terms of the equation—Maimon's point: by identifying, one asserts that one term underlies the other. In the reduction relation we preserve the identity while avoiding the symmetry, since in the general case there is no inverse reduction. In this move, important support is received from Frege, a comparison to whose position sets in relief various aspects of Maimon's doctrine.

A First Approximation

An initial approximation to Maimon's position can be attained through a comparison with Frege. Frege, like Maimon, tries to restrict the use of Kantian intuition, remaining faithful to the Leibnizian tradition. Nevertheless, there are two basic differences between the two positions. Frege, unlike Maimon, aimed his logicism at arithmetic without expanding it to geometry. Secondly, Frege proposed to dispense with intuition by reducing it to logic rather than to the theory of pure cognition.

But there are profounder differences. The *picturing* relationship in Maimon is, basically, different from a Fregean reduction. One may therefore presume that their respective conceptions of reduction are quite disparate. Following are three characteristics of Maimon's position not shared by Frege's. They lead us to the core of Maimon's conception.

First, Maimon is not a realist when this is understood as acknowledging existence outside of consciousness: "When I say that I am conscious of something, I do not understand this something to be a thing existing outside of consciousness—which would be a contradiction—but merely the unique mode of the consciousness, in other words, the special manner of the act itself" (*Versuch* 30). The objects of mathematics do not have an independent existence. In this respect, Maimon's position is different from that of Frege, and from such Leibnizians as Eberhard, his contemporary. Although the very claim that the sensible is a picture of the conceptual is redolent of Platonism, Maimon repudiates realism. We shall see later how Maimon uses his concept *picture* in order to define his notion of object and to account for the idea of existence outside consciousness.

Second, any attempt to explicate Maimon's position must refer to the complex relation between intuition and what underlies it. There is no a priori concept independent of, or purified independently of, intuition; in fact, the reverse is true: intuition supports, as it were, whatever it is founded on. This point is connected to the earlier one. Thus, in explaining the difference between his view and that of the transcendental realist, Maimon states: "He presupposes the objects [of the conceptual world] as fixed in and of themselves, whereas I presuppose them merely as ideas, or as objects which are not fixed in and of themselves, and can be thought of as fixed only by means of and within the sensation (like the differentials by means of their integrals). If an intuitional mode were eliminated, there would not be any intuition nor any objects of thought which are determined in and of themselves" (*Versuch* 206). The pure forms given to us through intuition in effect emerge from it. Intuition is a ladder on which we climb up to the forms—a ladder which *we do not* kick away when we have reached the top. Intuition is always in the background, for we always have to go back to it in order to progress further up the ladder.

This brings us to the third principle unique to Maimon's doctrine: the process of deciphering the conceptual element takes place gradually, in time. As will be explained below, Maimon does not presuppose the infinite understanding as an absolute reality that we gradually approach. Altogether, the idea of an infinite understanding is self-contradictory; it is actually a projection of the process of dismantling intuition to its ultimate components. Going up and down the ladder, which occurs within history, is what creates the very rungs of the ladder. In this respect, Maimon's ladder is different from Jacob's ladder, which presupposes heaven and earth, between which it mediates.[7]

Any explication of the picturing relation according to Maimon must incorporate the above principles.

Maimonic Reduction

We perceive a collection of propositions as a collection of sentences in a particular language, hereafter referred to as a theory. The theories discussed here are not necessarily closed under logical inference; that is, they are not each a collection of propositions that necessarily includes everything that follows from it. Likewise, we demand that the theories not be complete. A theory is complete if every statement formulated in the language of the theory is either provable or its negation is provable. In other words, anyone who knows the theory can infer from it every true statement about the concepts it contains. These two requirements are not usually made in the context of reduction, but they are not adopted to arbitrarily restrict the discussion. In truth, almost all interesting mathematical theories, such as arithmetic, topology, and set theory, are not complete, and we never possess the knowledge of all that follows from the axioms we have.

In addition to the above requirements, the language should be regarded as open: the assumption that we possess all the concepts is unfounded. It is not only that many definable concepts are yet to be defined, but, what is more, there is no reason to assume that all the concepts we will ever have could possibly be defined by the concepts we already have. In other words, the language can be regarded as essentially open, in the sense that the collection of concepts exceeds the closure of the concepts we know.

Let *T(t)* stand for the theory that meets the above requirements and which represents the state of our mathematical knowledge at a given point in time. Mathematical knowledge grows in two different ways. Either we enlarge the scope of our knowledge by adding a new truth, which has been derived (by proof) from *T(t),* including the addition of a new axiom, or else we effect a reduction of one or more intuitional representations to one or more non-intuitional concepts. In the later case we "effect an increase of the level of the intensity of the knowledge" (*Streiferein* 7). Let us designate the theory thereby obtained as $T(t+1)$. The second kind of procedure seems a natural candidate for the task of "dismantling intuition to its elements."

Whenever we perform a reduction of one concept or group of concepts to other concepts, we demand that the reduction be adequate. That is, we demand that every proposition we know to be true at a given stage t and which employs a concept based on a priori cognition shall be matched, by means of the reduction, with a provable proposition based on pure cognition. The crucial Maimonic warrant for such a step is given in his words: "If we survey all the theorems concerning a straight line, we will find that they follow not from its straightness, but from its being the shortest" (*Versuch* 68).

It should be emphasized that at no point are we effecting a comprehensive reduction of all the intuitional notions. In fact, the idea that the dissembling of intuition is infinite and can never be completed entails that even a single reduction of one concept is never complete. For the only thing one can state after proposing such a reduction is that the reduction of the concept will be relative to the state of our knowledge at a given time. It is quite possible—not necessarily by logical derivation from our knowledge at time t—that a new intuitional proposition will one day be discovered employing our concept, which will affect the truth value of the pure proposition. It is also possible that one will be able to prove, within the embedding theory, a statement that contradicts a basic intuition about the concept being defined. These possibilities exist even though we do not believe that they might be realized, just as the notion that scientific theories are in principle refutable does not contradict our belief in their validity. In this manner, mathematics gradually becomes more and more conceptual. Thus we preserve Maimon's third principle, according to which escape from intuition takes place in time.

What is conspicuous about the Maimonic notion of the reduction of one concept to another is the fact that the connection between the two concepts is asserted as a proposition in the very theory, rather than being a mere functional correspondence.[8] Herein lies the crucial difference between Maimonic reduction and a correspondence between two theories: Any two theories that involve objects with the same second-order properties can be made (that is, defined) to correspond to each other. Thus, the possibility of reducing any theory to set theory is simply the possibility of giving an interpretation to any consistent collection of second-order properties. On the basis of the above explanation, such reductions are just a kind of Hilbertization of the theory and an application of Goedel's completeness theorem: first we go to second-order properties, and then we simply find a model. This use of the concept of reduction expands it to the point of futility, for we are left with a situation in which, simply, too many things may be the underlying foundations of many other things. When Maimon proposes that the straight line be regarded as equivalent to the shortest line, he points out that one must preserve the conclusions that require the concept *straight line,* and that all true statements about a straight line are derived from its being the shortest. *But the proposition "the straight line is the shortest line" is evidently true.* It is not just an accidental transposition of the property "straight line" to the property "the shortest line."

We encounter a similar situation in the case of Frege's definition of the concept of number. The proposition "the number of elements in *A* is equal to the number of elements in *B* if there is a one-to-one correspondence between *A* and *B*" is a true proposition. This is an intuitive truth, which cannot be proved within set theory. However, it can be converted into a definition. I so construe Maimon's dictum that the understanding converts a reflective concept into an object (*Versuch* 68). Thus, in a deductive system in which we define concepts, every definition is also a theorem. However, in Maimon, the move emphasized is the reverse: a collection of concepts, or intuitive knowledge is converted into a definition. This is the concept-forming activity of the understanding, which is quite other than deductive derivation from axioms.

We are now in a position to provide a satisfactory definition of the basis of the claim that the essence of a *particular* concept is granted by a different *particular* concept. Without such a definition, we would be quite justified in asking what right Maimon has to argue that the concept perceived by intuition is equivalent to the concept on which intuition is founded. Before proposing the paradigm case of the straight line, as outlined above, Maimon adds that one should make sure that the substitute concept is equivalent to the concept being replaced from the point of view of its conclusions. When one concept is suggested as a substitute for another concept, it must satisfy the same propositions; that is, interchanging the concepts within the same propositions will preserve the truth-value of these propositions (forgetting for the moment the issue of modal relations, intentionals, etc.). Maimon lays down identity conditions for concepts: "two concepts with the same conclusions are identical concepts" (*Logik* 39). This is the principle of identity of indiscernibles, applied to concepts, which was acknowledged by both Leibniz and Kant. On the basis of this comment of Maimon's, the property granted to the intuition "straight line" is identical to that expressed in the reflective concept "being the shortest." The identity between "straight line" and the property "the shortest" is an identity that follows from a mathematical inquiry and is not analytical or vacuous. It is therefore clear why Maimonic reduction constitutes a disclosing of essence and is not merely the result of transposing one concept into another.

The above discussion calls for the introduction of certain obvious definitions, which would help us clarify Maimon's position. Reduction is a function from one set of concepts to another set of concepts, preserving the truth-value of the propositions, which we know to be true or false.[9] When we reduce one field to another (a field being a set of concepts and statements referring to a particular universe of objects), the field into which we effect the reduction is called the *embedding field,* and the field we reduce is called the *embedded field.*

Definition 1: Complete Reduction

Given two domains of knowledge, and given certain theories about these domains and the reduction of one theory to the other, we shall call the reduction complete if for every true statement in the embedded field (a priori cognition, in our case), there is a corresponding statement provable within the embedding field (pure cognition). One can define the complete reduction of a concept in a similar fashion. The reduction of one concept to another is complete if for every *true* statement in which the concept appears, there is a corresponding statement, determined by the image of the concepts being reduced, which can be proved in the embedding field.

One can distinguish different levels of completeness. One reduction is more complete than another if the set of truths that can be proved from the one includes the set of truths that can be proved from the other reduction.

Definition 2: Sound Reduction

A reduction is sound if every provable statement from the embedding theory (pure cognition) translates into a true statement in the embedded field. One should be careful to distinguish between completeness and soundness. A reduction is complete if for every true statement (in the embedded field) either the statement or its negation can be proved in the embedding field. A reduction is sound if for every provable statement in the embedding field there is a corresponding true statement.

Definition 3: Founded Reduction

A reduction of one concept to another is founded if the connection between the concepts is determined in a proposition that is presumed to be true. That this connection is obtained is not generally proved.

The requirement usually emphasized in the literature on reduction is that reductions are conservative; that is, everything that can be proved in the embedded field should also be provable within the embedding field. This requirement is different from the requirement of soundness. As noted, the indifference to epistemological considerations may have served to blur the distinction between that which is true and that which can be proved. As a result, the distinctions that, in my view, are essential to an understanding of Maimon's view, such as being sound, founded, and complete, have been largely ignored.[10]

Maimonic reduction is the founded reduction of an a priori concept to a pure concept. This foundedness is the basis of the claim that the reduction is sound, and if the embedding theory is complete, we have reason to assume that the reduction is also complete. The more true propositions we can prove about the embedded concept, the more we are able to rid ourselves of intuition, the closer we come to solving the quid juris question as regards those propositions that have been proved. The completeness and the soundness of the reduction become clearer as our knowledge expands. In other words, this reduction is confirmable, refutable, and founded. These features, one should note, can also be found in the interesting reductions found in mathematics and other sciences. This kind of reduction extends our knowledge in the domain that has been reduced; for it follows from the soundness requirement that it is sufficient to prove a theorem in the embedding theory in order to know a new proposition in the old theory. It may also be the case that there are no other means of knowing the truth-value of this proposition.[11] Hence, reduction can add to knowledge, and not only conserve it, as is customarily emphasized. It is the very fact that reduction enriches (rather than reduces) knowledge that requires that we add the requirement of foundedness and not content ourselves with a part of our truths being merely translated into some other part.

Examples

The conclusions of Maimon's response to the quid juris question are closely related to the mathematical developments of his time. The erosion of the special status of geometry amongst mathematicians began as early as the seventeenth century, and geometry gradually gave way to arithmetical methods and algebra. Euler applauds the algebraic method of analysis, Lagrange praises the abstract method of calculus and demands that we avoid drawing diagrams altogether, and Laplace remarks on the fertility of the analytical method as opposed to geometry. Maimon was aware of these developments, and he exhibited a profound understanding of their implications, which he eventually expressed in his critique of Kant.

The examples provided here are indispensable supplements. They enable us to concretize Maimon's conception—to lend it palpable form—so as to clarify aspects that abstract discussion keeps obscure, and to give concrete confirmation to his position, which will make plain just how remote his analysis of mathematics is from Kant's.

Defining a Function by Means of an Equation

The case of trigonometric functions highlights certain important features of the idea that mathematics is an activity that reveals the essence of intuition. The functions of sine and cosine are initially defined as relations between the sides of a right-angled triangle.

From this point on, we gradually free ourselves of intuition in two mutually complementary ways. At first, we extend the range of the definition of the trigonometric functions for the right angle and the zero angle. In both these cases, we go beyond the image of a triangle, which was necessary in order to define the functions of sine and cosine as initial values, because in a right-angled triangle there cannot be another right angle or a zero angle.[12]

In order to be able to apply theorems in cases where the functions are not well defined, we enlarge the range of the definition of the functions in accordance with the theorems about these functions. In the case of the zero angle and the right angle, we make use of the continuity of trigonometric functions and determine their value accordingly. We then proceed to expand these functions, in accordance with our basic knowledge of trigonometry, to an angle of a particular size.

However, this does not yet constitute qualitative progress. This expansion, like the previous one, is based on a reduction of angles larger than a right angle to angles smaller than a right angle, and, to the extent that the latter are involved, the value of their function is determined with the aid of intuition. Expansions enlarge the range of our concepts beyond our intuitions, but at this point they do not reveal the essence of these intuitions themselves, for the expansions are based on the definition of the sine and the cosine for angles smaller than the right angle.

Alongside the process of expanding the range of the definition, connections are revealed between the sine function and the cosine function, and these connections are formulated by means of the laws of arithmetic and algebra. One should emphasize that in order to discover these conceptual connections, we make use of intuitions and geometric constructions. The relation between these connections and expansions is not a simple one: on the one hand, when we extend the range of the definition of a function, we try to preserve those laws that have been established as valid for the reduced field. The laws thus determine the expansion. Yet, on the other hand, in surveying all the laws in question, we select those that will determine the expansion uniquely. In this way, it is our interest in the expansion that determines which laws will be preserved.

We now reach the critical stage: the conceptual connections revealed with the aid of intuition, which usually take the form of an equation, uniquely determine the functions of sine and cosine. In other words, one can prove that any pair of functions that instantiates these properties are identical to the pair of sine and cosine. We in effect fashion these conceptual connections into the definition of the functions of sine and cosine. The definition is effected by means of concepts that do not refer in any sense to the intuition of a right-angled triangle. Every value of these functions is determined by the equations that express algebraic relations. We perceive this definition (Maimon is no Platonist!) as capturing the complete essence of the trigonometric functions. At the end of the process, the functions of sine and cosine are mutually defined as the only functions that fulfill the following requirements:

1. $G(x + y) = G(x)F(y) + G(y)F(x)$
$F(x + y) = F(x)F(y) - G(x)G(y)$
2. F and G are continuous
3. $G(x) > 0$ for some positive neighborhood of 0
4. The minimal positive number x such that
$F(x) = 0$ and $G(x) = 0$ is 2π

Any G and F that meet these requirements are, respectively, the functions of sine and cosine.[13]

Maimon was not aware of this definition, but it most adequately represents his position as described in the previous section. Using a formulation adjusted to the citation at the beginning of this section, we can argue that the understanding converts a reflective rule, namely, 1–3, into a definition of a function whose picture we hitherto perceived through intuition. The cosine and sine functions we started out with, which belonged to the second level, are the pictures of the cosine and the sine given to us by the pure laws. But this requires an important reservation.

As stated, these definitions do not by any means appeal to the intuition of a triangle. They determine two functions by arithmetic operations —addition and subtraction and multiplication—as defined in the first stipulation, and by reference to *continuity*, as specified in the second postulate. The first postulate clearly belongs to the first level; but it should be noted that we have not completely freed ourselves from intuition, since the second postulate does appeal to intuition. This drawing upon intuition is immeasurably less than that required for calculating each value of the trigonometric functions, but nevertheless, as I shall elaborate later, it has not completely vanished.

It may be helpful to closely examine the implications of this point for the manner in which our judgments are determined. Before we have a definition, we compute the cosine of an angle—let us say, for example, of the straight angle, in the following way: we notice that the closer the hypotenuse approaches one of the sides creating with it an angle approaching zero, the more the length of the hypotenuse tends to equal the side of that angle. We therefore assert that the sine of a straight angle is one. But this proposition is based on intuition for we see, albeit apodictically, that the length of the hypotenuse gets ever closer to that of the side. But, its essence having been revealed, this proposition becomes one that is founded on a judgment that requires not intuition but arithmetic calculation.

This consequence is in accord with the Maimonic process of reduction. Through this definition, all the values for the functions of sine and cosine are specified systematically. This is no longer a function whose values are individually calculated, partly in an intuitive and partly even in an a posteriori manner, such as by computer (for one cannot calculate all the values of this function by a priori geometrical operations). The law of the function dictates its particularization, or the entirety of its values. This progression is, in Maimon's words, "a heightening of intensity—that is, a heightening of the unity of the concept and not merely an increase in the number of proven propositions. When we reach the essential level, we are able to formulate the connections between concepts that at the intuitive level seemed quite disparate. Conversely, when we reach a point in our inquiry where we are able to connect seemingly unrelated concepts, we have a reliable sign that we are approaching something essential."

Everyone would readily agree with me regarding the systematic sciences, that it is possible to reduce all the truths to a single truth. But it might be objected: What is the connection between the proposition "the air is resilient" and "the magnet attracts iron," and how is this related to Pythagoras's theorem? But what does this show? Only that we do not perceive these connections; and that is because we do not know the inner essence of the objects. If we discover the attributes of air, magnetism, etc., so as to be able to define these objects through their inner essence, these connections will readily transpire (*Versuch* 429).

The Conceptualization of Geometry

In the section "Symbolic Cognition" of the appendix to the *Versuch,* Maimon tackles the dispute over the relation between geometry and calculus. Arguing with Bendavid, Maimon defends the advantages of analysis over geometry (*Versuch* 276), explaining that we rarely employ construction in calculus. Maimon is alluding here to the mathematician's and physicist's habit of proving theorems about magnitudes and their interrelations by examining the relations between their differentials. These relations are transferred to the integrals—which can be perceived—of these differentials (*Versuch* 196).

Developments in the field of calculus and related theories provide a wealth of examples of the process of converting intuitive a priori cognition into pure cognition. One should note that these transformations are based on the definition of the concept of the limit, which replaced the intuitive-metaphorical definition that relied on notions like "tending toward" and "approaching." The conventional definition does employ concepts that still lean somewhat on intuition, but only minimally. By means of the concept of a limit, we can define a great many concepts, such as continuity, smoothness, area, length, and so on, and can turn propositions that use these concepts from propositions based on intuition into proved propositions.

We know intuitively that if one point on a line is below a straight horizontal line, and a different point on the line is above the straight line, then there is at least one point at which the line in question meets the horizontal straight line. This proposition, which is intuitively perceived, is Roll's theorem, which can be proved in basic analysis. But the progress in question is not only from judgments based on intuition to judgments that are less so. A similar process occurs in the case of concepts. Taylor's theorem allows us to present a rich range of functions, including all those that can be imagined as infinite polynomials, and to deal with them using algebraic methods. This is a generalization of the phenomenon that I described above, concerning the sine and cosine, to all intuitive concepts. Similarly, topology allows us to prove theorems at a more general level, which, without such proof, are merely intuitive. Jordan's theorem, which generalizes from the intuition that a prisoner cannot escape from jail without breaking the walls, is only one of many examples.

In these examples, Maimon would argue that the new definitions do not constitute a mere change of the language or a mere formal reconstruction. Maimon is of the opinion that calculus reveals those concepts our intuitions imitate and picture. We perceive the fact that we can prove intuitive propositions by means of these definitions as evidence of the definition's coherency, which reduces the reliance on intuition. Another approach to the conceptualization of geometry can be found in Felix Klein's work, which is one of the most thorough studies in modern geometry. According to Klein, geometry is the study of the invariants of particular groups. Every group of transformations has a theory that studies the invariants of this set—that is, the properties preserved by all members of the set. Klein presented the transformations that give us Euclidian, affine, projective, and other geometries and reached the following generalization: geometry is the study of the invariants of groups of transformations.[14]

One should note that although the whole study could be rendered completely formal by doing away with the intuition of the plane and focusing on the algebra of groups, Klein retained the intuition of the plane. The positivists did not like this alliance with the intuitive because they were trying to disassociate geometry from intuition (Reichenbach 1958, 100). The nature of Klein's work indicates his interest in the intuitive plane.[15] The transformations effected by the plane on itself are simply a conceptual expression of the way in which we perceive the objects on the plane, for they determine the concept of the indiscernibility of configurations in geometric space. The various different geometries are simply all the types of indiscernibility of geometric configurations relative to different languages. The richer the language the indiscernibility is relative to, the more concrete the geometry in question.

From Maimon's point of view, Klein's work is highly significant: the study of Euclidean geometry is just the study of a group (and as such is a pure algebraic study) that preserves size and angle. We have thus managed to expand the conceptual element without forgoing intuition.

The Continuum

Of all mathematical concepts, the continuum is, undoubtedly, very central, and the riddles pertaining to it are as compelling as ever. The definition of continuity by means of the concept of the limit, the definition of the real numbers by Dedekind, Descartes' algebraization of geometry—all these are, doubtless, important steps toward the conceptualization of linear continuity, and of the continuity of the Euclidean plane. Nevertheless, one must carefully distinguish between their contributions to the conceptualization of the continuum.

For example, the possibility of placing natural numbers on a continuum and then assuming the rational numbers, negative numbers, those which can be constructed and perhaps even all the real numbers, establishes a connection between two objects. One of these objects is perceived as the set of natural numbers, the other as a line. But one must be careful to distinguish between this record of points on a straight line and the conceptualization of the straight line, or of the property of continuity, just as one must distinguish between assigning an identifying number to each person and claiming that the series of identifying numbers is a group of people. Descartes' algebraization, for all its importance as a step toward the conceptualization of geometry, does not constitute a mathematical inquiry into the concept of continuity. The possibility of describing Euclidean lines by means of equations establishes an important and central connection between the two fields, but this does not amount to a conceptualization of the intuition.

Cantor's study of set theory changed matters. His great discovery, by his own evidence, is not the finding of a variety of infinities (as some philosophers are inclined to think), but the narrowing of the conceptual gap between the discrete and the continuous.[16] To achieve his characterization, Cantor uses the notions of order, density, and completeness. These notions are defined through attributes that also apply to infinite sets, which are also related to the picture of the straight line, or the plane. Thus, due to research in set theory, a position close to Maimon's becomes possible: the continuum is not a mere intuition, nor is it a pure concept quite independent of intuition. It is rather an intuition, which is put together in a manner necessary to its constitution, with a conceptual component.

This conceptual component becomes increasingly larger, as the theory through which we are trying to take apart the intuition of the continuum advances. Furthermore, this construction is not closed: the relation between the conceptual and intuitive components of our notion of the continuum is dynamic. One could say, generally, that in the course of mathematical history the continuum concept underwent an interesting transformation. The conceptual component increases at the expense of the intuitive.

Euclidean geometry and Greek analysis restored to intuition at each step. In Descartes' work, and in Newton's and Leibniz's analysis, appeal to intuition diminishes in favor of the conceptual. Weierstrass and Cauchy continued in this vein very successfully. Nowadays we appeal to our intuition of the continuum only in order to establish the connection between real numbers and the straight line. In proving propositions we do not consult our picture of the continuum, since the picture provides us with no theorems about the continuum. In fact, the picture does not enable us to distinguish between the line of real numbers and the set of rational points on that line. In proofs of propositions about the continuum we use Cantor's definition. The intuitive straight line is perceived through this lens, hence through set theory, which is presupposed by Cantor's definition. But, as set theory is incomplete, the manner whereby we apprehend the continuum is also incomplete. This is manifested by our inability to decide every question regarding the continuum.

As hinted in the preceding section, one should resist the temptation to get rid of intuition altogether. One may think that Cantor's definition of real numbers is the definition of the straight line, without this involving any theory. In other words, one may think that the set of real numbers is, at the outset, or by stipulation, the intuitive straight line and not a theory about the intuitive straight line. Mathematicians interchange *R* —the set of real numbers—with the straight line, supposedly demonstrating that the equivalence between them is not a matter for consideration or the conclusion of a particular theory, which may turn out to be wrong. Others may regard intuition as a constituent of the context of discovery, owing to which constituent we are able to discover the definition of *R*, whereupon we may ignore it.[17] Maimon, by contrast, would argue that Cantor's definition of the continuum is nothing but a theory of the continuum that appears in intuition.

Mathematicians' habit of failing to mention the complexity of the relations between the intuitive straight line and the real numbers is, so Maimon would argue, a consequence of the desire for brevity and the wish to avoid repeating this *presupposition* over and over again. The correspondence between real numbers and the straight line is not a trivial identity, guaranteed in advance by stipulation or by virtue of a simple definition.

This point also enables us to clearly see how it is that reduction constitutes an exposure of the essence. When, in the course of developing a theory at the pure level, we come across a conclusion that does not accord with its being a theory about a field of cognition at the a priori level, we try to narrow the gap. The mathematician, as a revealer of the essence, seeks to adjust propositions from the pure level to propositions from the a priori level. There are two ways in which he can do this: by changing our intuition or by changing the theory on which it is founded. In the interpretation, adaptations of this kind explain Maimon's position, according to which intuition is an image of the conceptual.[18]

Kant's dualism of understanding and sensation, which is echoed in the dualism of concept and intuition, has to be modified. Judgment that is aided by intuition is replaced by judgment that requires less intuition. Maimon shows that this transition is possible through his idea of reduction. This reading frees us from the schematic formulation that is bound with Maimon's metaphorical wording and reveals his position within the framework of the development of mathematics.

At this point many questions can be raised about the validity of Maimon's solution. Specifically, what Kant would have replied to Maimon and what implications Maimon's reduction has beyond this dialogue. Perhaps the most interesting question is about the relation between Maimon's critique of Kant's dualism and modern discussions around the myth of the Given. We should not forget, however, that quid juris is not only Maimon's challenge to Kant but also his way of presenting his whole system. As we develop Maimon's ideas, we remain with the following questions:

1. What happens outside mathematics?
2. If we accept Maimon's criticism of Kant's dualism, what are the amendments required in the Kantian system with the idea of the thing-in-itself, the concept of the object, the ideality of space and time, the idea of affection, and the transcendental deduction?

4

An Interlude

Polarization

MAIMON, LIKE Kant, drew on mathematics only to the extent that it could contribute to an understanding of explicitly epistemological or metaphysical questions. The Maimonic reduction carries within it a rationalist, anti-Kantian kernel, which finds its expression in positing a concept of objectivity that is stronger than Kant's. Thanks to this conception of objectivity, Maimon can renew the possibility of metaphysics and, at the same time, show that the thrust of the Humean skepticism has not been removed from it. The general course is therefore a critique of Kantian transcendentalism: Maimon places the Kantian transcendental in between the rationalist and the empiricist poles. I have labeled this move the Maimonic polarization.

Objectivity

Kant's two principle conclusions—interactive dualism and the renunciation of the absolute and objective validity of the representation of space—arise in the course of his analysis of the possibility of geometry. To the extent that intuition was considered to be autonomous rather than conceptual, as followed from the analysis of geometric theorems, their connection to the objective was no longer tenable. In contrast, for Maimon, intuition is in itself conceptual, though of a different state; it does not merely correspond to, or embody, the concept, but is actually grounded in it. On the basis of this notion, Maimon considers himself justified in demanding the extension of the objective validity of geometric propositions.

For the sake of precision on this important point, it is worth recalling the manner in which Kant establishes the objectivity of geometry: "For we cannot judge in regard to the intuitions of other thinking beings, whether they are bound by the same conditions as which limit our intuition, and which for us are universally valid. If we add to the concept of subject of a judgment the limitation under which the judgment is made, the judgment is then unconditionally valid. The proposition that all things are side by side in space, is valid under the limitation that these things are viewed as objects of our sensible intuition" (*Critique* A-27, B-43). For Kant, at least as far as one can conclude from the above passage, restricting the validity of a judgment to humanly accessible phenomena is what guarantees its objectivity. The proposition "All objects are beside each other in space" is not analytic, and if it quantifies over things in general, we cannot ascribe to it unconditionally. It is only when we limit the proposition to things that can appear to us that it becomes absolutely valid. As far as objects that are not accessible to us, or creatures that perceive them are concerned, there is nothing we can say. One way to paraphrase this idea is from within—the proposition is absolutely valid and the outside world is unlikely to present any problem as far as I am concerned, as it will never be available *to me.*

Maimon is not satisfied with this view, but not because of the association of space with the representation of space, as one may be led to believe if one interprets Kant as a psychologist. The source of Maimon's response is more complex, as transpires from the following passage:

> I, on the other hand, believe that these judgments, like all synthetic judgments, are only subjective truths, that is, they constitute, as far as I am concerned, a necessary mode of cognizing a certain object. These propositions, then, are not valid for objects in general, nor for this particular

> object from the point of view of all thinking beings. The proposition "every triangle is identical to itself," on the other hand, is an objective truth, for I think of a triangle as identical to itself in such a way as not only I myself, but any thinking being at all must think not only of a triangle but of every object as identical with itself. If this were not so, then no thought would be possible at all. Mathematical theorems, then, are objectively true, but only if one assumes the objectivity of their principles (which of course is possible). In any other respect, they are, just like the principles themselves, only subjectively true. But this does not make their usage any less correct, for just like their own truths, their application, too, is valid only as far as we are concerned. (*Versuch* 406)

A truth which is valid only as far as we are concerned is a subjective truth; a truth which every thinking being must necessarily accept is objective. The phrases "as far as I am concerned" and "subjective," in this context, are not necessarily meant in a psychological sense. Maimon is alluding to the distinction between the sensuous a priori and the pure a priori, and his argument is thus directed at the degree of objective validity of the Kantian a priori. In fact, elsewhere he states this point explicitly: "I would like to note here, for once and for all, that I consider the necessity, which Kant calls objective (a condition for the objective experience), as merely subjective. And this on two accounts: (a) Suppose that some synthetic law was necessary for granting objective reality to sensations. For example, we cognize sensations A and B by applying the law of causality; however, another thinking being may cognize the very same sensations by applying another law" (*Versuch* 418). The transcendental is a condition for the objective experience, not the objective in itself. In order to clarify this point, it is worth studying Maimon's definition of objective truth:

> A truth which has been acknowledged by a particular thinking being is, to this extent, merely a subjective truth. However, when it is acknowledged by him in such a manner as to render it necessarily recognizable by any thinking being, it is an objective truth. Our sensual intuitions, for example, insofar as they correspond to known forms, are merely subjective, as there may always be thinking beings with forms of intuition completely different from our own. Therefore, these very forms, although

they are a priori within us, have only subjective reality. The same applies to the forms of our thought, for there could always be thinking creatures that connect phenomena (if these are accessible to them) by means of completely different forms, thereby making them into objects. (*Versuch* 151)

Maimon distinguishes between the transcendental, in the sense of being a condition for the sensations and experiences captured in the synthetic propositions, and the objective. In contrast to the passage from Kant quoted at the beginning of this chapter, the transcendental is not a sufficient guarantee of objectivity. This is not so because *we* might discover it to be mistaken but because of the problems involved in the claim that it binds all intelligent beings. One should note that in voicing such a doubt, Maimon is not alluding to a world that exists somewhere out there, nor, of course, to the thing-in-itself. Maimon's idealism does not permit him to express his doubt by speculating on the connection between the transcendental and the reality beyond consciousness. His doubt concerns the existence of a consciousness with a different transcendental structure, displaying a different world. Maimon raises the possibility of the existence of worlds different from our own—worlds constructed by "beings who associate phenomena by means of completely different forms, thereby making them into objects."

This possibility reflects the autonomy of intuition within the Kantian position and is in fact a logical conclusion of this position. But here is a serious problem involved in this concept-intuition dualism: even if it does not amount to psychologism, it is liable to lead to relativism. From Maimon's perspective, although Kant holds that intuitions without concepts are blind, he keeps the possibility of associating the intuition with several different concepts. To return to Maimon's example, one would have to argue that what Kant eliminated was not the possibility of other creatures failing to associate the representation of the straight line with the concept of the shortest distance, but the possibility of them associating this representation with another concept, and not necessarily one compatible with straightness.

We are not dealing here with an attempt to refute Kant's solution to the problem of cognition, for Kant regarded the possibility of there being creatures whose apprehension is not space-time apprehension as a *conclusion* of his discussion. What Maimon is demanding is a response to the quid juris question in light of the Kantian solution. The basis for this demand is the discovery of pure cognition and the objectivity of logic. The degree of objective validity Kant managed to achieve, as far as geometry is concerned, is insufficient, and Maimon wants to solve the problem of cognition, or at least the problem of the connection between the pure and the a priori, without restricting the validity of our knowledge. Kant, Maimon's defenders may argue, underestimated the danger of relativism implicit in his arguments, the extent of which only transpired in later years. Indeed, it was Kant's transcendentalism that showed Nietzsche and Cassirer the way to using expressions such as "different concept-scheme," "world-constructing cultures," etc. One should say, in Maimon's favor, that he was the first to thoroughly comprehend this implication.[1]

Maimon's response to the quid juris question reduces the autonomy of a priori cognition by regarding it as a picture of the conceptual. Judgments based on intuition are actually conceptual judgments. This possibility provides the backing for Maimon's challenge to Kant. Kant cannot argue that Maimon's question about the connection between intuitions and concepts is meaningless, as he claimed in the letter he sent to Maimon. This course is blocked, as Maimon in fact managed to reply, however schematically, to his own question.

Maimon's response does not require realism; it proposes a way to counter relativism without attempting to separate truth from the subject. In the above definitions, Maimon explains that objective truth is a truth the cognition of which, on the part of the subject, involves the cognition that it binds all thinking beings. It is not a truth that describes a world independent of thinking beings, but rather one that is imposed on all thinking beings. But Maimon's response hangs on the assurance that we possess such objective cognition. One may ask how we know that logic does not just describe our own mode of thinking. What is there to guarantee the universality of pure cognition? Some of the remarks I made in the course of the analysis of pure cognition may help answer this question, although, at this point, one should note Maimon's caution: "It would appear," he explains, "that we do not in fact have any test for objective truth." It is difficult to distinguish, "from the inside," that which binds oneself, and binds other subjects, only in a problematic manner, from that which binds all subjects. Nor can we resort to certainty to help us out here, for this is involved in both cases. "Nevertheless," Maimon adds:

> If we consider this question more accurately, we must realize that this doubt does not invalidate our thought. Nobody can expand his doubt if he does not wish to contradict himself, as if he says that there may be thinking beings with forms completely different from our own, he must still admit that these thinking beings, if indeed they are such, must necessarily have something in common with us. . . . I do not, of course, wish to undertake to determine just what this common element consists in; I leave this to my opponent; in other words, it is he who must determine what he understands by the phrase "thinking being," and, when he has expressed his opinion on this matter, he shall find it necessary to acknowledge some known truths objective. (*Versuch* 152–53)

This suspicion that there may be other thinking beings who do not accept logic or pure reason puts the onus on anyone expressing such doubts to define what he means by a thinking being. For one thing, it obliges one to explain what one means by the notion of thought that is not subordinate to the laws of logic. Maimon puts the skeptical ball in the other court, but he is aware of the fact that this is not sufficient ground for a test or a criterion of objective truth.[2]

In light of this point, the significance of the Maimonic reduction is obvious: by means of such reduction, one can heighten the degree of objectivity of our knowledge. The objectivity of a particular kind of cognition is directly related to the range of subjects for which it is valid in terms of their cognitive apparatus. Reduction to pure cognition, or to logic, shows *me* that a cognition I thought of as valid only for myself, and perhaps for other beings with a sensory structure similar to my own, is in fact valid for a wider range of subjects. Maimon, in contrast to Kant, aspires to, and succeeds in, salvaging the certainty and the productivity of cognition without sacrificing its degree of objectivity.

The Maimonic reduction does not aim to increase my degree of certainty regarding the knowledge I possess. Its more important purpose is to extend the validity of this knowledge. Maimon thus raises an aspect of this issue which, to the best of my knowledge, has never been emphasized in discussions of Frege's logicism. In view of the phenomenon of paradoxes, modern philosophers such as Quine have rejected the claim that we can establish anything at all by the reduction to logic. But this very rejection presupposes equivalence between grounding something and increasing its degree of certainty. Maimon, and perhaps Frege too, did not need reduction in order to heighten our degree of certainty regarding various propositions. In fact, Maimon poses the quid juris question about mathematics only because he has subjective certainty. It is obvious to him that he will never be able to refute the claim that the straight line is the shortest distance between two points.[3] But there is a point to this subjugation: it is not dependent solely on the contingent structure of my own consciousness. The reduction constitutes an attempt to heighten the degree of objectivity of my knowledge, assuming that it is certain.

Skepticism

The value of these pursuits of objective validity in cases where certainty, on which the very beginning of this whole process of inquiry depends, is of special interest. In the case of mathematics, relatively speaking, there is a consensus about the certainty of its truths. In fact, the certainty and generality of mathematics led Kant to the conclusion that science is restricted to subjects with particular forms of sensation. Maimon acknowledges the dependency of geometrical truths on human cognition but expresses the hope that this dependency is not attached to a particular mode of cognition but, rather, is a consequence of cognition in general. But when we move away from the field of mathematics to other areas of cognition, we have no such certainty about our judgments, and we must ask ourselves whether in this case, too, Maimon's analysis is still valid.

Maimon does not discuss this issue explicitly, but it is possible to suggest two ways in which to answer this question. The first is to argue that as long as the quid facti question is not resolved with respect to a particular area of cognition, we may not begin the process of uncovering its objective component. What seems to support this approach is the notion that one cannot think of the foundation of anything unless this thing is factual. The analogy with the concept of explanation may be useful here, for we cannot explain an event that has not taken place. If we could come up with an explanation for a nonexistent phenomenon, say the boiling of water at 200°C, then it is not clear how we could fit it in with the explanation of why water boils at 100°C. If, for the purpose of the argument, we adopt the deductive model for explanation, it will follow that our theory provides a proof that water boils both at 200°C and at 100°C, which means that the explanatory theory contains a contradiction. One can find the grounds for this move, though not in exactly the same formulation, in Maimon's reluctance to pose the quid juris question with regard to causality.[4]

Another reason that justifies that approach derives from the fact that certainty is necessary in order to begin the process of the Maimonic reduction. The transition from the proposition "a straight line is the shortest distance between two points" to the definition "a straight line as the shortest distance between two points" is possible only if the proposition that serves as the basis for the given definition is certain. Otherwise, the proposition would not sustain the definition and the reduction would be ill founded, amounting to no more than a mere replication of terms. Things are even clearer if we take into account Maimon's idealism. If the world is constructed through consciousness, then in the absence of objects existing outside our consciousness, we will tend to rely upon certainty to sustain this world.

The second approach to the possibility of expanding the idea of Maimonic reduction outside mathematics seems to me just as interesting as the first. This approach involves the distinction between certainty and necessity—with both of them considered to be different from the a priori.[5] Bearing in mind this distinction, we can view Maimon as one who grounds objectivity in necessity rather than in certainty. Indeed, the difference between a priori and pure cognition is not on the certainty that accompanies both, but in the necessity, which we attach to the pure cognition but not to the mere a priori. Though we cannot base this distinction either on a realistic conception of the necessary or on its analysis in terms of possible worlds, this does not pose a serious limitation, since his paradigm is logic and pure cognition, which are based on the analysis of thought. A proposition or a synthesis is necessary if the components of the synthesis are interdependent and inextricable from each other, and not if they are real in every possible world.

By taking this approach we can modify the above analogy between explanation and the uncovering of the objective foundation. Instead of referring to a nonexistent phenomenon, the analogy now concerns an uncertain phenomenon. Explanations of uncertain phenomena abound in the science of physics, and, as a matter of fact, they increase the certainty of the phenomenon we are trying to explain because of the autonomous basis of the theories in which they are interwoven.

Nevertheless, when a phenomenon is not recognized as a fact, our search for objectivity can only result in a hypothetical objective validity. We cannot say, "If it is true that *P,* then *S* is the underlying objective concept of *P.*" As far as Maimon is concerned, we cannot say here, "If it is true that *P,* then *P* is objective," for he defines truth in terms of objectivity, and consequently the hypothetical proposition becomes void. We shall therefore have to say something like this:

1. As far as mathematics is concerned, it is possible to argue that *P* is certain and probably *S* is the objective element of which *P* is the image where *P* belongs to a priori cognition and *S* to the realm of pure cognition.
2. In physics and in relation to transcendental concepts—categories and forms of sensation—it is possible to argue that *P* is not certain and probably *S* is the foundation of *P.*

Without an underlying objective concept, *P* cannot be objective. Thus, either the given phenomenon *P* is a synthesis of the faculty of the imagination or it has an objective foundation. Now, if it is not a synthesis of the imagination, then *S* may be its foundation—where *S* belongs to the realm of pure cognition.

This disjunction, which is bound to emerge whenever we depart from mathematics, is what I designate the Maimonic polarization. This polarization operates in relation to the dimensions of space and time and with respect to causality and substance. The results of this operation provide us with the Maimonic method, which is characterized by an original combination of Hume and Leibniz. This combination cannot be understood apart from transcendental idealism and therefore it would be a mistake to think that Maimon combines the positions of Leibniz and Hume by adopting a Leibnizian theory in mathematics and a Humean theory outside mathematics. Such a formulation of Maimon's approach would be overly simplistic. The rationalistic aspect is not characterized by a return to a theocentric outlook, for Maimon's concept of objectivity makes no reference to God, but by the central place assigned to the subject. Similarly, the question that conveys his skepticism is not "How do I know that there is something external which corresponds to my representation?" Rather, it is "How do I know that the given syntheses—whether or not they involve certainty, and, furthermore, whether or not they are the condition for any sensation—have an objective foundation?"

This skepticism is completely different from that of Descartes or the skepticism Kant finds in Berkeley; hence, Kant's dismissal of skepticism is irrelevant to Maimon. Maimon's skepticism is also distinguished from the skepticism that was prevalent in his time.[6] He does not argue against such Kantian postulates as Platner's contention that the mind is incapable of investigating its own working, or Schulze's postulates, which are based on the doctrine that it is impossible to know the thing-in-itself. Maimon took upon himself to reply to skeptics like Schulze by differentiating between Schulze's dogmatic skepticism, which failed to comprehend the Copernican revolution, and his own critical skepticism.

Concerning this point, Rotenstreich (1968) observed that Maimon's skepticism escaped the problem of skepticism being turned against itself—an argument to which sweeping skeptical anti-Kantian positions were vulnerable. Secondly, and perhaps even more important, Maimon posits a model of irrefutable knowledge, and in this way he makes it possible to introduce standards that cannot be met by empirical knowledge. But even if we acknowledge the uniqueness of Maimon's skepticism, it can still confuse us. Beiser, who examines the critical nature of Maimon's skepticism, understands it as questioning the correspondence between the conceptual and the sensible. Consider, for example, the following passage: "Assuming that truth is not the correspondence of representations, the problem still arises of how such heterogeneous representations as the a priori concepts of understanding and the a priori concepts of sensibility even correspond to one another" (Beiser 1987, 292). Beiser identifies

Maimon's skeptic contention with the question of the correspondence between the conceptual and the sensible. Yet the skeptical question is even stronger: it poses the question of whether there is anything conceptual that corresponds to the syntheses produced by the faculty of imagination.

A subtler problem adheres in Rotenstreich's definition of Maimon's skepticism. Rotenstreich explains that Maimon's skepticism does not demolish the rationalistic ideal but only demonstrates the problem involved in realizing it. Rotenstreich ascribes Maimon's skepticism to our finite nature or, more precisely, to the notion that our finite understanding is incapable of producing from within itself the manifold of intuitions and of completely uncovering the rationalistic element underlying the cognition of the imagination. This view is problematic, since the finite potential of human consciousness is also employed in mathematics, with its heavy reliance on a priori intuitions, which also do not lend themselves to complete uncovering. If our finite mind were the reason for Maimon's skepticism, there would be no difference between his discussion of mathematics and his discussion of the natural sciences. Once again, Maimon's skepticism does not lie in the fact that we cannot uncover the objective foundation of the phenomenal world, but relates to the very question whether our scientific knowledge is indeed grounded on such a foundation.

This inaccuracy does not detract from the uniqueness Rotenstreich finds in Maimonic skepticism. It is true that one cannot use the skeptic argument to counter Maimon's skepticism, and Maimon did not destroy the possibility of the ideal. Nonetheless, his skepticism does not consist in the claim that it is not possible to realize the ideal. If such were his claim, it would not be any different from the neo-Kantian positions of Cohen, which regard man's fulfillment of his vocation as an infinite task. In fact, almost anyone who argues for the validity of the ideal shares the view that it is impossible to conform to it. Hence it is not clear what is so remarkable about ascribing to Maimon a skeptic position of this sort.

Maimon's skepticism consists in posing the question of whether the notion of a continuous uncovering of the essence of things is valid even if this pursuit is recognized as an ideal. From this it becomes clear that the uniqueness of Maimon's position, which combines skepticism about the existence of ideal and rationalism, is that Maimon showed that the ideal is not impossible as such. *Perhaps* we live in a world that exceeds the syntheses of the imagination, a world in which the syntheses of the imagination have an objective basis and all intelligent beings form part of the same reality. This is something that we shall never know for sure, and not because of our finite intelligence. What we do know, however, Maimon would add, is that Kant has not refuted this possibility. And as long as this possibility is in force, so is metaphysics.

The following statements are implied by the Maimonic reduction:

1. Following Kant, Maimon distinguishes between objectivity and being a condition for experience, or even a condition for sensation.
2. The reduction allows us to raise the objectivity of our cognition.
3. It follows from the mathematical discussion that for every concept we must seek the objective element that corresponds to it. This search is always hypothetical and never certain.
4. When we are not sure of the fact itself, as in causal propositions, we are still allowed to speculate on the objective foundation of these propositions.
5. In this case, however, we do not know whether *X* is indeed the objective foundation of *P*, nor do we know whether there is anything at all which is the objective foundation of *P*.
6. This leads to a special kind of skepticism, which is not encapsulated in questions such as "How do I know that there is something that corresponds to my representations?" or "How do we know that A is always followed by B?" Instead, the pertinent question is this: "How do I know that the syntheses of the imagination are valid for all thinking beings?"
7. Kant's transcendentalism does not dispense with skepticism, and even if we were to assume that objective experience does exist, Kant's explanation of its possibility is liable to result in relativism.

§ 5 §

Aesthetics
The Critique of Ideality

The first concepts to be subjected to the Maimonic polarization are space and time. Maimon is not satisfied with Kant's claim that these are transcendental concepts. He tries to reinstate the rationalist alternative while at the same time considering the possibility of adopting a skeptical position on space and time. All the same, it would seem as if, given these two options, Maimon is inclined to challenge Kant's position on space and time more from the rationalist position than from the skeptical one. In an attempt to implement this approach, Maimon searched for the concept of which space and time are the pictures: "The difference between Kant's theory and my own is, therefore, as follows: in Kant's opinion, space is merely a form of intuition, whereas in my opinion space, as a concept, is the form of all objects in general and, as an intuition, it is the picture of this form. Space, in his opinion, is nothing in the object itself, when it is abstracted from its representation. In my opinion, however, space is always something as far as any subject is concerned; it may indeed be a form, but it is a form grounded in the object" (*Versuch* 427).

The Theory

The fact that space is an intuition does not obviate the possibility that it may also be a concept. The characteristics proposed in Kant's metaphysical exposition of space, which are based on the notion that only one space is accessible to us and that space is a condition for any empirical intuition, are not sufficient to reduce the validity of the concept of space. This validity is extended in two ways: space applies to all objects and space is a form for subjects in general. The second part of this extension is clearer than the first, for here Maimon demonstrates his strict notion of objectivity. According to Maimon, the representation of space is valid not only for "humans and perhaps for other creatures," as Kant postulates, but also for all possible rational beings. This extension of validity is still consistent with Kant's notion of space as intuition.

The first part of the extension, namely that the representation of space is valid for all objects, requires explication. First, it is not clear why every object must have some relation to space, or that space is an integral form of any object whatsoever. Furthermore, this also goes against Maimon's notion of correlated or real objects, which are not situated in time or place. The only way to understand this is not to claim that every object relates to space, but that the form underlying space—its concept—applies in some way, which needs to be explained, to any object. Indeed, when we examine the concept suggested by Maimon as the foundation of space, we get a clearer picture of what he actually means: according to Maimon, the concept in which space is grounded is the notion of difference. When we perceive two bodies in space by means of our senses, we picture the fact of their being conceptually different from each other. The conceptual difference underlying the given objects is pictured, or generally perceived by our senses, as an occurrence at different points in space and time. Space applies to every object because the notions such as difference and identity apply to all objects.

This schematic presentation of Maimon's theory of space suffices to determine its relation to what we have described above as Maimonic reduction. Maimon's discussion of space immediately follows his analysis of the claim that "the straight line is the shortest path." In other words, Maimon's theory of space is completely analogous to his theory about the relation between a priori cognition and pure cognition. This analogy allows Maimon to concur with Kant that space and time are a priori intuitions but at the same time to claim that intuitive space is a picture of difference, in the same way that the straight line that appears to our intuition is a picture of the shortest path.

This passage enables us to clearly comprehend the way in which Maimon grants objective validity to a priori cognition. While Kant sought to establish the validity of geometry by presenting space as a subjective condition, Maimon believes that the degree of validity obtained by this course is insufficient. Human beings may indeed be creatures whose perceptions are bounded by space and time. As such, they cannot refute their mode of sensation by working from the inside, for to do so they would have to exceed their natural abilities. However, this should not stop us from an attempt to guarantee the absolute objectivity of space and time, that is, their validity for all thinking subjects. Thus, though space and time may not be things-in-themselves, existing independently of consciousness, this does not allow us to conclude that they are not objective in the strict sense: "I therefore agree with Kant that space, if regarded as an intuition in itself (but not as the picture of a relation) has a merely subjective reality, and that things that we see as occurring in space may not appear as such to other beings; but I would add the point that this subjective phenomenon must have objective grounds—grounds which, being objective, are necessarily cognized equally by all thinking subjects" (*Versuch* 182).

An example from physics may be helpful. The way in which one perceives a yellow light is not necessarily the way in which it would be perceived by other beings. A being with a different sensory system may respond differently to the light, but one should not equate the light itself with the way in which it is perceived. It would be more accurate to say that this is the subjective way in which we perceive light, whereas, objectively, light is electromagnetic radiation and, as such, exists independently of the way it appears to us.

According to Maimon, space is a backdrop against which we place objects, which are conceptually different from each other. Any conceptual difference is perceived as a difference in position against the backdrop. This theory should not be confused with similar conceptions, according to which space is the condition for identity between objects.

Kant, for example, thought that space is not abstracted from experience but is a condition for discernment and separation. "For in order that certain sensations be referred to something outside me (that is, to something in another region in space from that in which I find myself), and similarly in order that I may be able to represent them outside and alongside one another, and accordingly as not only different, but as in different places, the representation of space must be presupposed" (B-38). Maimon, however, adopts Kant's theory in order to defend the thesis that space and time are objective. It is not their occupying different places that is responsible for their difference, but on the contrary, their conceptual difference results in their having different places in space.

It was this motivation—not an insight or a deduction—that led Maimon to make the principle of the identity of indiscernibles into a regulative rule: Given the occurrence of two objects that are identical in all their properties, we can conclude that we know these objects only partially (*Logik* 179). As in the case of mathematics, here, too, an intuitive truth—that objects that occupy different places are usually different—acts as a means to support the reduction of the representation of space to the representation of conceptual difference. This may not be sufficient to establish the completeness of the reduction—which, in this context, means the impossibility of discovering identical objects in different positions in space. But this reservation does not invalidate Maimon's proposal. By the same token, there is no guarantee that we will not one day discover a theorem that will force us to reject the equation of being a straight line with being the shortest distance between two points. But until such a rejection is proposed, we are quite entitled to regard the property of being the shortest distance as an objective foundation for the intuition of a straight line.

Notwithstanding the specific contents, and even the validity, of Maimon's ramified conception of time and space as outlined in the above discussion, what is crucial is the role it plays in his argument against Kant. Maimon finds it necessary to argue that Kant's discussion on synthetic knowledge is not sufficient to restrict either the metaphysical interest or our requirement for rigorous objectivity. The Leibnizian proposal that space is related to conceptual difference is only one way of illustrating that Kant's notions fail to do away with the rationalist ideal. Indeed, Maimon's general move can be implemented in other ways, which are not necessarily Leibnizian. Any theory that would explain that time and space derive from conceptual relations and that they are not independent variables may serve as a substitute for Maimon's proposal, in the sense that it has the same critical function vis-à-vis Kant.[1]

A Short Comparison with Frege

Frege and Maimon are two Leibnizian, post-Kantian philosophers who oppose transcendental idealism, hence their connection. Yet they criticize Kant's restriction of the objectivity of space and time from two opposite angles: Maimon is a radical idealist while Frege is a realist. Nevertheless, the two philosophers use a reduction to a pure theory in order to attack Kant's thesis on intuition. In fact, there is a striking similarity between Frege and Maimon on this issue, as can be seen from a Fregean remark like this: "Furthermore, the example shows us how pure thought, without any reference to content derived from the senses or from a prior intuition, can create, from content derived only from its own structure, judgments which may apparently be founded on reference to intuition. This can be compared to the action of compressing air, whereby the air, which to a child's mind is nothing, can be turned into a visible drop of water" (*Begriffsshrift* 23). Frege's above description is surprisingly similar to the way in which Maimon describes the relation between intuition and its conceptual basis. A priori intuition is classed together with the senses, while the human intellect prior to the reduction appears here metaphorically as a young child. Moreover, for both of them intuition can be thought of as a concept in a different state of aggregation. Intuition is not autonomous; rather, it is a limited mode of perception, analogous to that of a child, of a truth that transcends sensibility.

Other similarities are found in more familiar passages from Frege's works: "I understand objectivity to mean what is independent of our sensation, intuition and imagination, and of all construction of mental pictures out of memories of earlier sensations, but not what is independent of reason—for what are things independent of reason? To answer that would be as much as to judge without judging, or to wash the fur without wetting it" (1950, 36).

Finally, here is an enlightening excerpt from Frege's *Foundations of Arithmetic:* "We can say, almost in words familiar to everybody: the object of the inquiry of reason is reason itself. In arithmetic we do not investigate objects that we know from the outside through the senses, but objects that are given to our understanding as its next of kin. These are objects whose transparency with regard to reason is absolute" (1950, 115).

These excerpts suggest that in essential questions about the nature of number and the nature of objectivity and its relation to rationality, Maimon and Frege belong to the same philosophical heritage. Yet the marked difference between Frege and Maimon lies in ontology. According to Frege, idealism leads to psychologism and to doubts about the objectivity of human cognition. Frege defended the objectivity of mathematics by drawing on logicism and realism, while Maimon suggests the possibility of defending objectivity by combining a position that resembles logicism and an idealistic position.[2] This gives Maimon one clear advantage, as he can ground the objectivity of geometry and mathematics, while at the same time suggesting a theory of sensibility that Frege does not supply.

Frege attributes to humans the ability to perceive objects and objective truths that are not carried by them. But if human beings are flesh and blood, equipped with intuitions, memories and sensations that have nothing to do with objectivity, as Frege himself admits in several places, then how can this bundle of occurrences and fragments pave its way to the essence of things? With this question we do not deny the fact that human beings perceive truths, but how this process is accomplished is unintelligible to us. In Maimonic terms, we do not deny the quid facti question, but we lack the answer to the quid juris question. Frege insinuates that intuitions play a role in the discovery context, but in fact, this is not true as far as arithmetic is concerned. As for geometry, Frege espouses the Kantian analysis that in order to establish geometric truths we need to resort to a priori intuition.[3] Now this invites the following question: How can one adopt the notion of a priori intuition without acknowledging idealism? Conversely, if the use of intuition in arithmetic is suspect because it leads us to psychologism, why, in geometry, does it not restrict the objective validity of our judgments?

Dummett, indeed, stresses the importance of this question: "In speaking about space, as described by geometry, we are not, for Frege, speaking of our intuitions or sensations. If this leaves it obscure why we should treat our intuitions as a ground of knowledge of geometry, that is a lacuna that Frege might have filled had he ever written positively about geometry" (Dummett 1991, 156). This question, however, is not confined to geometry; it has to do with the general question about the relationship between subjective sensations and intuitions, which belong to one's inner world, and the objective propositions, which are based on them yet are not dependent on them.[4]

Maimon utilizes the idea that intuition is not added to the concept, but it is another phase of the concept, a metaphor that Frege accepts in the case of arithmetic. This allows him to raise the level of objectivity of geometry and to still be able to offer an explanation for its accessibility to our cognition.

The emphasis here is on materialistic realism. If Frege had argued for a Platonic position, which claims the physical world is an expression of a conceptual world, and that our cognition has some affinity with this world of concepts, then he could have avoided the idealism of Maimon and held on to realism. The challenge that remains in this case is to account for the possibility of the connection between one's intuitions and the theorems of that world of concepts, which requires that he admit to the possibility of a connection between understanding and intuitions.[5]

The sharp distinction Frege likes to make with regard to Kant, between intuitions or representations, on the one hand, and concepts and objects on the other, relegating the former to the psychological space only, prohibits him from giving them a place of importance when they serve us in the apprehension of nonanalytic truths. This comparison with Frege indicates clearly where the important advantage of Maimon's idealism lies: it gives a place for intuition, and in this it agrees with Kant, and questions the level of objectivity Kant had managed to give geometry without taking a realistic approach. The latter approach guarantees objectivity, but in its way channels sensibility and everything else that's human to a position of insignificance. Maimon's achievement is made possible by extending the space of logical cognition into pure cognition, and by pointing out that a number of propositions based in the a priori sensual cognition represent pure truths. Frege cannot deny the validity of this line of thought, as he himself had used a similar line in his foundation of arithmetic, claiming on the basis of the reducibility of *some* of the arithmetical truths to logic that arithmetic is probably logic.[6]

The Skeptic Move

Maimon criticizes Kant's position on time and space from the skeptic viewpoint as well. In taking this step, he is not looking for the rational framework of time and space but questions Kant's very claim that these are a priori forms of our sensibility and that everything we perceive appears to us in time and space. Maimon is also apprehensive "lest at some point in time I might perceive some object also as not existing in space (or in time) and hence we have no ground for raising the generality of representations—which was obtained a posteriori, by way of induction—to the level of a priori necessity" (*Versuch* 342).

Maimon's question does not relate to the possibility of perceiving in relation to time and place while ascribing to them properties different from those given in the geometric theorems. His doubt is much more radical: How do I know that every perception must appear to me in time and space? In this way, Maimon questions Kant's presupposition that "space is not an empirical concept abstracted from external experiments" (*Critique* A-23, B-38). By raising this doubt Maimon suggests the possibility that time and space are derived from experience and that one day we may perceive things outside time and space.

Kant has put forward an argument based on the apodictic nature of the geometric theorems. If space were an experiential representation, so the argument goes, we would have to regard the geometric theorems, too, as derived from experience, but this cannot possibly be true. This argument did not convince Maimon, who would have countered that such an argument only moves the problem one step backward, for we have no certainty that geometric theorems are true. Indeed, whoever is willing to raise such a radical doubt about the possibility of sensations removed from time and space will not find it difficult to suggest the possibility of sensations related to a non-Euclidean space. This notion was not at all foreign to Maimon, who anticipated the possibility of non-Euclidean geometries.

Apparently, Kant answers Maimon by his "Metaphysical Exposition," according to which space is not derived from experience because in abstracting anything from objects in the external world, space is already presupposed. The very act of abstracting requires that we presuppose objects outside of us and then extract from them some common property. But in order for this process of abstracting to occur, the objects must be located at different points in time and space. This argument is discussed by Kant's commentators (see, for example, Allison 1983, 81–114), yet, regardless of the details of this argument, it is difficult to see how it can convince anyone who shares Maimon's doubt. Suppose that this act of abstraction is conditional upon the representation of space. Why, then, should this serve as a proof that perception outside time and space is not possible?[7] Maimon confirms his own doubt by drawing a comparison with the rules of logic: "Here things are not the same as in a contradiction, in relation to which we are convinced that it cannot be thought of, because we recognize contradiction by certain [formal] signs, without having to determine what they must refer to. Here we acknowledge that so far, we have not experienced an intuition divorced from time and space.[8] There, one acknowledges the impossibility, whereas here we are only unaware of the possibility. Kant merely presupposes the fact, without actually proving it" (*Versuch* 340).

The very comparison between the rules of logic and perceptions bound to space is interesting and important. A contradiction emerges from the pertinent signs regardless of the particular content associated with them. The signs suggest that a contradiction is something that can be detected without thinking; the rules of logic are embodied in the very notation. In contrast, geometric principles, or the restriction of sensation to time and space, are not transparent to us.

Presumably, one can answer Maimon's question by defining sensibility or perception as the process by which the subject situates phenomena and objects in space and time. But by doing so we constrain the meaning of sensibility in such a way that its aspect of necessity is not manifest. Maimon, like Kant and others, believes that sensibility is essentially a conjunction of multiplicity, a passive state of reception, and, generally speaking, he dwells on its abstract properties, which can be defined by concepts that make no reference to time and space. In particular, Kant's or Maimon's notion of sensibility does not refer to the specific sensory apparatus of humans, for both Kant and Maimon are of the opinion that there can be creatures who do not perceive phenomena in relation to time and space. The statement that sensibility is immanently connected with time and space commits us to more than what Kant and Maimon regarded as essential to sensation.

Nevertheless, if one examines the issue more carefully, one can see that in his skepticism about the a priori nature of time and space Maimon confuses two different questions. The first one is whether sensibility removed from space and time is possible, and the second is whether the concept of space is derived from experience. Indeed, even if there is a third form of sensation, it does not follow from this that space is a concept abstracted from experience. The fact that as a form of sensation time is different from space, as well as the fact that our inner life is not situated in space, does not make space a matter of experience.

The methodology employed in Kant's system allows us to come up with a third form of sensibility, without thereby making space empirical. This possibility depends on the question of whether we possess some other synthetic knowledge, which cannot be accounted for by the forms of sensation accessible to us. According to the interpretation that I presented in the first chapter, viewing space as a form of sensibility is a conclusion that emerged, at least partially, from Kant's solution to the question about the possibility of geometry. An analogous step also led to the notion that time, too, is a form of sensation, which accounts for the synthetic judgments in kinetics.

It follows that the notion that time and space are forms of sensibility is a conclusion that does not derive from the definition of sensibility. The logic behind the Kantian analysis leads to the hypothesis that if there were a science that generated synthetic propositions and the object of this science were neither space nor time, then it would be possible to expand the form of sensibility in such a way that it would incorporate this science within its range. Thus, the question about the possibility of some other form of perception targets on the possibility that there might be a group of synthetic a priori propositions not currently familiar to us. Therefore, Kant did not raise the question about the possibility of a third form of sensibility. Maimon does not share this view. It is against this difference of opinion that the concluding line in the previous citation should be read; Kant has not proved that the sciences have reached their end. According to Maimon, we need to consider even the possibility of further developments in the science of formal logic. It is possible, then, that there is a third sensual form, but this is not enough to prove that the representation of space is inductive.

Monism

Kant's analysis of mathematics has led us to the ideality of space and time and to concept-intuition dualism, both of which form the basis of transcendental idealism. These two positions are based on, or are at least closely affiliated with, the existence of synthetic a priori judgments. Maimon abolishes the absoluteness of synthetic judgments, thus reopening the question of ideality; he then goes on to examine the various dualisms involved in the *Critique*. Maimonic reduction does not entail the conclusions that intuition is superfluous, that there are no synthetic judgments, or that one can simply know things-in-themselves. Kant, so Maimon would argue, revealed the existence of synthetic judgments and the dualism of concepts and intuitions. But Kant's system cannot be complete as long as we accept these distinctions as absolute. This criticism of Kant is now presented by Maimon as an interpretation of Kant, for "Kant presupposes the reality of synthetic judgments (in mathematics) only from the point of view of our limited intellect, and on this point," adds Maimon, "I would readily agree with him" (*Versuch* 62).

Maimon admits that were we describing cognition at a particular point in time, Kant's account would be faultless. However, he argues that as soon as we allow ourselves to resort to gradual succession across time, Kant's distinction becomes blurry. Any judgment based on an a priori intuition is, potentially, a judgment based on a concept. It is only from our point of view that sensibility and understanding are considered to be two different things. From the viewpoint of infinite understanding, however, sensibility is but incomplete understanding (*Versuch* 183). We see, then, how abolishing the sharp distinction between synthetic sentences and sentences based on a concept entails the abolition of the distinction between concept and intuition. Kant, of course, did distinguish different degrees of subjectivity in syntheses, but he regarded the intuition required for a given judgment as essential to that judgment. He could not accept the possibility that judgments based on a priori intuition, like geometrical theorems, will become, by means of a process akin to that described by Maimon, judgments not based on intuition to the same extent.

Maimon's monism can be illustrated by means of the balloon metaphor, which we introduced in the first chapter of this work in order to capture the idea of Kant's form of sensation and a priori intuition. In Maimon's view, the claim that geometry consists of describing the surface of the balloon is the result of ignorance as to the source of the truth and validity of our judgments. The further we get toward understanding the grounds for synthetic judgments, the less synthetic they become. The finite understanding cannot apprehend an infinite number of concepts at once and thus has to resort to the imagination, which crams an infinite number of concepts into the picture, and it is this cramming that forms the walls of the balloon. The development of science is the dissection of intuitions into their components, and it is this process that reconstructs the concepts that make up the balloon.

An analogous process takes place at the objective level. The less dependent on intuitions our judgments are, the more general and independent of us they become. In terms of the balloon metaphor, one could say, somewhat simplistically, that the balloon's surface gradually becomes transparent, thereby revealing the concepts pressing against it. This makes them less restricted to residents of the balloon, and the things that can be seen through it are eventually recognized as necessary by everybody. At the point of limit the balloon disappears. Yet we can never actually reach this limit, for were it possible to do so, we would be completely liberated from our finite constitution.

The distinction between intuitions and concepts is a dynamic one. There is no intuition without concepts and no concept without intuitions, but there are two ideas that can be extrapolated from both of them: infinite understanding and sensation. These are absolute, limiting concepts, which have no independent existence. Cognition, or reality, falls within the range that stretches between these ideas.[9]

Maimon discusses the implications of this picture for the concept of intuition. As he writes in his short survey in the introduction to the *Versuch:* "A sensation is a mode of the faculty of cognition, realized in it merely in a passive manner (without any action). However, the sensation which is passively realized is only an idea, which we can always approach by restricting our consciousness, but can never reach, as lack of consciousness = 0. Therefore it cannot be the mode of the faculty of cognition. Intuition is the mode of the faculty of cognition which is actualized partly in a passive way and partly in an active way" (*Versuch* 168).

Pure passivity exists only as an idea. There are no sensations completely free of conceptual processing. Pure consciousness or experience is only possible if accompanied by a concept. Sensations are the result of abstraction and, in fact, there are no subjective sensations and no subjective experiences at all; every experience involves a certain degree of objectivity, and the more conceptual the experience, the more objective it becomes. All the rules we recognize as objective are based upon concepts and not upon intuitions. On the other hand, unorganized multiplicity of sensations is not possible, and the notion of it is simply a matter of extrapolation.

This extrapolation is conducted in the following manner: We first base a particular judgment on intuition. By means of a Maimonic reduction we then become aware of the fact that the same judgment can be established with less reliance on intuition. This may lead us to conclude that the intuition could appear independently of the concept with which it is connected in the judgment. But this would not be a valid conclusion, just as one cannot validly conclude from the fact that one can apprehend both straight and bent lines that one can apprehend lines that are neither straight nor bent. Maimon does explain that we can reduce the conceptual component through which we apprehend something, but the notion of being able to continue this process ad infinitum, to the point of complete multiplicity, is only an idea.

Analogously, one cannot conclude from the fact that for every number there is a smaller number that there exists such a thing as the smallest number. Infinite understanding, too, is only an idea. It functions as an ultimate, limiting term, which enables us to analyze the ongoing process of cognition: "This continuous 'approaching' is made possible by the constant separation and generalization of concepts and judgments, whereby they remove themselves from matter and keep approaching form, although its complete attainment is not only a mere idea but is even self-contradictory for it is both an object and not an object" (*Versuch* 242).

Maimon regards the process of historical development as the process of gradual distancing from matter and approaching form. The correlate propositions deal with form devoid of matter. The development of knowledge consists in an ever-increasing exposure of the essence on which intuition is founded. In terms of the previous discussion, it leads us away from the second level, which involves a priori intuition, toward the first level of pure concepts.

Maimon's extreme idealism, which assigns a major role to the notion of consciousness, makes it quite clear that the claim that our understanding is a limited version of the infinite understanding is not to be understood as a definition of human understanding. Spinoza saw man as a modus of God, God being the object that precedes man both in nature and in concept. Maimon's conception is different. The infinite understanding is a fiction, in reference to which we interpret what human understanding has to complete in order to solve the quid juris question. The concept of infinite understanding should be understood from the concept of finite understanding: the schema of the infinite understanding is our own understanding.

The infinite understanding is not an object. It is not even a special kind of object, in which all of reality is involved. It is a concept of limit—an idea. As Maimon hints, only the members of a series, which by their succession gradually approach the limit, actually exist; the last number is not one of the members within reality. In the same manner, pure sensation is only an idea. Our cognition, so Maimon explains, begins and ends in the middle. And as Maimon is an idealist who makes no distinction between cognition and the world, it thus follows that the world, too, begins and ends in the middle. In other words, the world is constantly in the process of becoming.

Maimon's idea that intuition is the picture of the concept, which was clarified in the Maimonic reduction, allows him to answer the quid juris question, as we did in the third chapter. Furthermore, it opens the way for him to amend Kant's analysis of synthetic a priori knowledge. Intuition becomes conceptual in the process whereby propositions based on certain judgments gradually become based on pure concepts. The fact that something is an intuition, even an a priori intuition, does not entail that there is no concept lying at its base. Maimon extended this realization to space by suggesting that the ideas of difference and identity are the concepts on which space is founded. These are general concepts, and they originate with pure a priori cognition, but every rich and interesting concept from pure cognition that fits the general qualities of a priori intuition is a natural candidate for the foundation of the intuitions of space and time. Kant's analysis did not disprove the possibility that at the base of time and space there is a pure mathematical structure, and that the placement of things in space and time is the human way of grasping this structure. The ideality of space and time and Kant's interactive dualism need adjustment.

6

Reconstructing the Outside

KANT REFUSED to write a letter of recommendation for Maimon's book, but this was only, writes Kant, "since it is after all largely directed *against me.*" Indeed, Kant admitted, "none of my critics understood me and the main question as well as Mr. Maimon does." He regarded Maimon as an original philosopher who took a very different path than his. In summing up the difference between his position and Maimon's, Kant wrote:

> If I have correctly grasped the sense of his work, the intention is to prove that if the understanding is to have a law-giving relationship to sensible intuition (not only to the empirical but also to the a priori sort), then the understanding must itself be the originator not only of sensible forms but even of the material intuition. . . . Otherwise the *quid juris?* cannot be answered adequately. . . . An a priori synthesis can have objective validity only because the divine understanding, of which ours is only a part, is one with our understanding. . . . I, on the other hand, . . . ascribe to the understanding the synthetic unity of apperception, through which alone the manifold of intuition, . . . in a unified consciousness is brought to the representation of an object. (Kant to Marcus Herz, 26 May 1789, *Letters* 151–52)

Kant, like almost all of Maimon's commentators, was not aware of the role of philosophy of mathematics in Maimon's system. He therefore could not trace the argument that starts with Maimon's quid juris question in mathematics, which was answered by his notion of reduction and concludes with Maimon's monism.

Affection and the Thing-in-Itself

Although Kant tries to construct our objective world from materials accessible only to humans, there are still a few external remnants he was reluctant to renounce, which are reflected in some of the key concepts of the Kantian system—that is, the concept of the thing-in-itself and, more important, the concepts of affection and representation. In spite of the idealism of his system, Kant avoided the claim that we are responsible for the creation of the sensuous representations that appear to us, ascribing their source rather to the affection of the thing-in-itself. It is in these tenets of his system that Kant expressed his avoidance of the idea of intellectual intuition. In contrast, Maimon tries to construct his entire universe without resorting at all to external entities. In this respect, Maimon is not different from many of Kant's successors, who equally objected to the assumption of the thing-in-itself. In fact, it is widely agreed that it was Maimon who paved the way for the relinquishing of this assumption.[1]

Nevertheless, there are unique aspects to the course that Maimon takes in order to avoid the thing-in-itself, and these merit a closer discussion. Indeed, Kant's theory of the thing-in-itself is not a mere supplement that can be abandoned without further explanation. At this point, one should recall the systematic structure of the Kantian position: even if the thing-in-itself first arises from the awareness that we do not create our own sensations, it gradually comes to play a more crucial role, as we have seen in the context of the theory of the ideality of space and time, and as later appears in the form of a hypothesis that enables us to salvage freedom. Furthermore, the key terms of the Kantian system, the concept of affection and the concept of representation, still maintain some connection with the thing-in-itself. One cannot let go of the thing-in-itself without first examining the implications of such a step for the key terms in the Kantian system.

Maimon succeeds in avoiding the premise of the thing-in-itself along with all external remnants of cognition. His monism and version of the ideality of space lead him to eschew the thing-in-itself, but also to reclarify certain terms in the Kantian system that still imply some reference to the outside.

Affection is one of the most difficult concepts in Kant's system. This concept, whose correlate is activity, or spontaneity, is conceptually associated with the notion of sensibility. "Intuition exists only in so far as an object is given to us, while the object is a possibility, at least as far as humans are concerned, only if it somehow affects the mind. The capacity for receiving representations through the mode in which we are affected by the objects is called 'sensibility.' By means of sensibility objects are given to us, . . . by the understanding they are thought, and from it arises conception" (*Critique* B-33). This characterization however expresses and even comes immediately after Kant's dualism of sensibility and understanding (Critique A-15, B-29). Maimon's rejection of the theory of the two stems of human consciousness leads him to revise the Kantian postulate and to reclarify the notion of affection. He draws a distinction between different concepts of affection:

> There are three ways in which we are affected by sensibility:
> a) By not having any consciousness of the concepts in it;
> b) In such a way that even from the point of view of the concepts we can attain, we must necessarily cling to sensibility, in order to become conscious of them;
> c) By means of thus obtaining these concepts themselves and the relations between them, usually in an incomplete manner and over time, and in accordance with the laws of sensibility. (*Versuch* 183)

We find a similar distinction in Maimonides's *Guide of the Perplexed,* in the course of his explanation of "the difference between the agent's knowledge of his own acts and an outsider's knowledge of these acts" (484–85). In this sense, understanding or spontaneity forms part of the Maker's knowledge, whereas by affection one gets to know the product merely from the outside. Maimon, in keeping with his monism, imparts an epistemic twist to this distinction: the finished product matches the sensuous intuition. In the first sense of being affected, Maimon associates our passivity regarding the given with the fact that we do not know the concepts involved in the sensuous component that forms a part of every cognition. The relation of affection, which was previously perceived as a relation between the thing-in-itself and the mind, is thereby transposed to the human consciousness: being affected is a mode of unknowing. The clearer we grasp the conceptual element of the given, the less we are affected.

Maimon does not define what an external object is, but he defines what it means to have a consciousness of an external object.[2] The justification for his doing so is clear, for there is no external object outside consciousness. Thus, although Maimon rejects the notion of entities outside consciousness as incoherent, he is not satisfied with Berkeley's view that *esse* is *percipi*. The notion of the Maimonic reduction suggests another relation: When I make a particular judgment, I am conscious of the fact that by doing so I am not aware of the conceptual component that will be manifest in other judgments. Our imagination translates this unawareness into something that appears out there, outside consciousness.

It is noteworthy that Maimon was not the first to translate a theological idea into an epistemic one; Kant made a similar move before him. We learn this from the addition that "the object is not a possibility, at least as far as *humans* [my emphasis] are concerned," which is found in the previous excerpt from Kant (*Critique* B-33). Even the notion of another kind of understanding, from which these two stems of cognition derive, alludes to Spinoza: when we renounce Berkeley's notion of matter, the Cartesian-Spinozian extension is replaced by sensibility, and the notion of various attributes that are issued from God is replaced by the idea of the monism of understanding and sensation. Hence, it is clear why Kant characterized Maimon's position as Spinozian (Kant to Marcus Hertz, 26 May 1789, *Letters* 152).

One can now assess Maimon's critique of the thing-in-itself. As opposed to the accounts by Bergman (1967) and Atlas (1964), Maimon does not offer his explication of the thing-in-itself merely as a response to the problems involved in this concept. This problematic aspect of

Kant's work is not the be-all and end-all of Maimon's philosophy, and it is certainly not its point of departure. In particular, Maimon is not concerned with Jacobi's question of how the concept of causality can be applied to the thing-in-itself. He is mainly concerned with exploring his quid juris question and providing an answer to this question. His clarification of the relation of affection follows as a natural outcome of this answer.

Indeed, the critique of the concept of the thing-in-itself is already embodied in the critique of the ideality of space, for the concept of the thing-in-itself appears as a necessary component of Kant's argument that space and time are ideal. The assertion that space is ideal is tantamount to the claim that space does not describe any property of the things-in-themselves. From the moment we criticize this assertion and define space as the picture of identity and difference, or of some other pure concept, we cannot share Kant's view that space does not reflect the relations between things-in-themselves. Maimon is not attempting to find faults in Kant's conclusions about the thing-in-itself—and in this respect he differs from Jacobi—nor does he counter Kant's assertions. Rather, he reverts to the reason that compelled Kant to presuppose the thing-in-itself and to the source that motivated him to affirm its existence. More specifically, restricting the validity of geometry is one way of cognizing the thing-in-itself. Kant writes: "Therefore understanding restricts sensibility without thereby extending its own field and by alerting sensibility not to pretend that it applies to things in themselves, but only to phenomena, it cognizes the thing in itself" (*Critique* A-288, B-344). By dispensing with this restriction and the dualism that is implied by it, Maimon gets rid of one mode in which the thing-in-itself is thought by us.

The second path by which Kant arrives at the notion of the thing-in-itself is the argument that the thing-in-itself is the correlate of sensibility. Our faculty of imagination does not have any control over our sensibility, and we therefore have to regard the mind as being affected by the thing-in-itself. The term *correlate of sensibility* refers to the reality that we can only regard as the basis for, or source of, sensuous intuitions.

Maimon's solution to the quid juris question involves the assumption that intuitions are pictures of concepts. This assumption, which was originally employed for a different purpose, achieves the purpose for which Kant first proposed the thing-in-itself, for intuition now has a support, or a foundation, which is *not* an extra-cognitive reality.[3] Maimon invalidates the reasoning that led Kant to the concept of the thing-in-itself. Again, he is not raising a question internal to Kant's system about the relation between the thing-in-itself and sensibility. By thoroughly investigating the problem of the relation between intuitions and concepts and the implications of regarding synthetic judgments as merely a limitation of our understanding, Maimon manages to avoid the thing-in-itself. In fact, his account of the concept of affection makes the thing-in-itself, as the affection of sensibility, redundant. We may continue to talk of objects affecting us, but we do not understand this to be a kind of affection by something outside our consciousness: "This illusion can be described as follows: the representations of the objects of intuition in time and space are, as it were, images produced in the mirror (the empirical ego) by the transcendental subject of all representations (the pure ego); however, the images appear to come from something behind the mirror (from objects different from ourselves) . . . The expression 'outside of us,' however, need not deceive us as if something stood in a spatial relation to us, since space itself is only a form within us. However, this 'outside us' means only something whose representation does not involve any activity of which we are aware" (*Versuch* 203). On the basis of this view he asks: "what makes all of these [philosophers] presuppose a transcendental object (of which they know absolutely nothing), or what makes them attribute our consciousness to something outside consciousness (as suggested by the word 'representation')?" (*Versuch* 161).

From Representation to Consciousness

Maimon believes that every philosophical system must begin with an undefined concept, and he sees the highest-class concept as consciousness. Maimon's position is thus directed chiefly against Kant, according to whom representation *(Vorstellung)* is the highest determinable. Indeed, in a letter to Kant, Maimon explains that the concept of representation led philosophers astray as it prompted them to invent an objective substratum for each act of the mind (20 September 1791, *Letters* 176).

The centrality of the concept of representation within the Kantian system is undisputed. This notion is treated as a key concept within the Kantian system by Kant's commentators and critics alike. Reinhold, a commentator and disseminator of Kant's work, emphasized the importance of this concept during Kant's lifetime. His commentary, which consists mainly of organizing and elucidating the grounds for the Kantian system, treats the distinction between the represented and the representing as the basic tension underlying the definition of cognition. The concept of representation later became a favorite target for realistic assaults on Kant coming from philosophers such as Bolzano and Frege. But whereas this line of criticism endeavored to defend the notion of an object outside of consciousness, which has no connection with the subject's states of mind as implied in the concept of representation, Maimon's critique was completely opposite. The concept of representation alludes to something outside itself; the concept of consciousness, however, does not entail such an allusion; it is a determination of the subject and herein lies its advantage.

According to Maimon, there is no definition of the concept consciousness, and, indeed, it is impossible to provide such a definition once the concept is considered the highest-genus concept. In lieu of such a definition, Maimon states that "it is an obvious and well-known fact that consciousness asserts itself to be the highest-genus concept" (*Logik* 245). However, this should not lead one to conclude that Maimon makes no attempt whatsoever to establish the primacy of this concept.

In order to elucidate the Maimonic reduction, we assumed Maimon's extreme idealism, which rejects the notion of something outside of consciousness as incoherent. But by the realization of this approach in mathematical practice, we make this first step somewhat more reasonable. Maimon's response to the quid juris question leads to a certain conception of affection, according to which it is not the case that the object is located outside the medium of human cognition, from where it affects our mind. If affection does not imply something outside of cognition, then one must reevaluate the notion associated with the concept of representation, according to which a representation stands for something outside the field of cognition.

The concept of consciousness subdivides into two types: representation and embodiment or realization *(Darstellung)*. In neither case are we referring to anything outside our consciousness:

> When I say: "I am conscious of something," I do not understand this something to be a thing which exists outside consciousness, which would entail a contradiction, but merely the special mode of consciousness . . . The term representation, as employed by simple consciousness, is misleading in this context, since in fact it is not at all a representation, in the sense of presenting that which is not as that which is; rather, it is actually a realization *(Darstellung)*, that is, the representation of something which did not exist, as such, beforehand. A representation, in the strictest sense, is a repetition of part of a synthesis relative to this synthesis. The consciousness of each part of the synthesis before one becomes conscious of the synthesis, is *not* a representation but a realization, as it does not refer to something outside itself. Likewise, the complete consciousness of all elements of the synthesis, and thus of the synthesis itself, is not a representation but the realization of the (intellectual) thing itself. (*Versuch* 349; my emphasis)

Sensations, colors, pain, and imaginary pictures are not, in and of themselves, representations, for they do not involve any reference to other syntheses.[4] Similarly, one could say that from the point of view of the infinite understanding, which embodies the synthesis that constitutes the object, there is no representation. Neither newborn babies nor God ever represent, for their perceptions are not made in relation to other things. A representation is a modification of consciousness, characterized by the fact that it refers to an object—but this should not lead one to conclude that there is an object outside of consciousness. Outsideness is the appearance of the affection, which, in turn, is simply the manner by which something with an unknown conceptual element appears to the mind.

In a letter to Kant, Maimon remarks on the implications of this point for the concept of intuition: "An intuition, in my opinion, is not related to anything other than itself. It becomes a representation only by being united with other intuitions in a synthetic unity, and it is as an element of the synthesis that the intuition relates itself to the representation (that is, to its object). The determined synthesis to which the representation is related is the represented object" (Maimon to Kant, 20 September 1791, *Letters* 176). An intuition is a representation when it is perceived as part of a larger synthesis—otherwise, it is realization. Maimon connects his theory of the object as a synthesis with the distinction between representation and embodiment. Since the object is a synthesis of intuitions and concepts, or a synthesis of concepts formed in accordance with the law of determinability (see chap. 7), the reference to an object by our cognition can be perceived as the relation of the part to the whole.

Nevertheless, this should not lead one to conclude that being part of a whole is tantamount to being a representation. No single component of a synthesis is in itself a representation; the operation by which a given component is ascribed to the greater whole is what constitutes a representation. Accordingly, a patch of color is either a mental state or a representation: in and of itself, it is a mental state, but in relation to something else, it is a representation.

Transcendental Deduction

Maimon's critique of Kant would be incomplete were it not possible to extract from it his position on Kant's transcendental deduction. In the absence of such an argument, Maimon may appear to be trying to catch Kant out on the issue of schematism without dealing with what constitutes the heart of Kant's innovative approach—namely, the proof that the categories apply to experience. Certainly, when we recall Maimon's skepticism about causality and his criticism of the inference of the categories from logic, the demand for a response to the question of transcendental deduction becomes somewhat less urgent. But all the same, it would seem that in avoiding the issue of deduction, Maimon would be doing an injustice to the principles of the Kantian approach. For even if the categories are actually not those laid down by Kant, and even if we set Maimonic skepticism aside for a moment, Maimon still has to account for the connection, so crucial to Kant, between the possibility of self-consciousness and the possibility of a law-governed world. This connection does not presuppose any specific categorical form, and the question of its possibility arises if we temporarily abandon Humean skepticism—which even Maimon is occasionally willing to do.[5]

A schematic description of the Kantian account of deduction is sufficient to demonstrate how Maimon would interject his criticism into the Kantian argument. As mentioned, according to Kant, if the manifold that appears to be intuition is to be regarded as my own, it must be accompanied by the representation of "I think." This consists simply in the consciousness of a necessary synthesis of these representations by means of what Kant calls "original synthetic unity of the apperception." The only syntheses available to the understanding are generated by means of the categories, which are derived from logic by the metaphysical deduction, and which constitute the understanding's only modes of connection. This should lead to the conclusion that the representations conjoined by the "I think," which result from the constraint that my intuitions can be ascribed to a single consciousness, must be connected by the categories.

This general account of one of the finest gems in the history of philosophy will serve us in defining Maimon's position on the deduction. And, indeed, the *Versuch* is full of comments highly relevant to this issue. For example, one can invoke Maimon's views on the "Self," his critique of the Kantian distinction between the "phenomenal self" and "the self as a thing-in-itself," and, consequently, the application of Maimon's critique of the concept of representation to the very notion of "the representation of the 'I think'"[6] (*Versuch* 209–10). A more central aspect of the *Versuch,* relevant to the issue in question, can be found in Maimon's attempt to connect the possibility of personal identity with logic and syntheses in accordance with the law of determinability (*Versuch* 155–59). These and other of Maimon's ideas, which merit a separate discussion, are undoubtedly relevant to Kant's deduction, and they can be essentially understood in light of Maimon's general position.

In truth, Kant himself can be of some assistance here. Immediately after introducing the idea of the "I think," which accompanies every representation, Kant alludes to the junction at which Maimon would probably have interjected:

> [F]or the reason alone that I can comprehend the variety of my representations in one consciousness, do I call them "my representations" [. . .] Synthetic unity of the manifold in intuitions, as given a priori, is therefore the foundation of the identity of apperception itself, which antecedes a priori all determinate thought. But the conjunction of representations into a conception is not to be found in the objects themselves, nor can it be, as it were, borrowed from them and taken up into the understanding by perception, but it is on the contrary an operation of the understanding itself, which is nothing more than the faculty of conjoining a priori, and of bringing the variety of given representations under the unity of apperception. This principle is the highest in all human cognition. (*Critique* B-135)

The final words of this passage suggest the context for Maimon's interjection, for according to Maimon, human cognition is not essentially different from the cognition of the infinite understanding, or, more precisely, it is a schema that enables us to apprehend the infinite understanding. Indeed, as Kant goes on to explain, from the perspective of the infinite understanding—or of the "intuitive understanding," as he called Maimon's system—transcendental deduction would look completely different. This understanding, "in and through the act itself of its own self-consciousness, contains the manifold in intuition" (*Critique* B-135). From the perspective of the human understanding, the situation is quite different. The manifold comes from outside the understanding, which conjoins it, thus allowing for self-consciousness. The case of the intuitive understanding is quite a different matter: the self-consciousness of the infinite understanding is not a supplement to the way the understanding connects the manifold, and is not a product at all. The only product involved in it is the manifold itself. Thus, self-consciousness, which in humans is a derived product, is the source of the intuitive understanding, and the multiplicity, which is given in the case of human consciousness, is a product from the perspective of the intuitive understanding.

Transcendental deduction refers to human cognition, which has to assume the multiplicity of sensibility as a given outside of the understanding. This clearly shows that Maimon's quid juris question does not have to wait for schematism and already comes up in response to Kant's transcendental deduction. Maimon cannot accept as intelligible the principal story of the deduction, according to which the understanding conjoins the intuitive manifold in accordance with its own forms. In other words, the deduction constitutes an attempt to prove that the categories apply to experience, while acknowledging the possibility of the synthesis of the "I think" with the multitude of data, but this account leaves one with another problem: the intelligibility of the connection between different and independent factors.

Kant returns to this issue in a long footnote in the middle of his account of the deduction (*Critique* B-144–46). He emphasizes that the whole deduction argument assumes that it is necessary for the manifold of intuitions to be available even before the understanding performs the synthesis, and independently of it (B-145). He also infers from this point that, for the divine understanding, the whole function of the categories is redundant. Obviously, transcendental deduction is not valid for anyone who does not share Kant's premises as to the structure of human cognition. Indeed, Kant realized that Maimon's criticism touched the very heart of his account of deduction, as indicated by the fact that his above-mentioned footnote is repeated word for word in his letter of reply to Maimon.

For Kant, this divine understanding has no place in the description of human cognition. More precisely, it is merely a negative concept, introduced in order to comprehend the idea of human understanding. When we say that this is an understanding, which, by the very representation of objects also creates the objects, we are defining it by negating an essential property of our understanding. Without this negative characterization, we would have no conception of such a divine understanding. For Maimon, however, we comprehend the divine understanding by regarding the human understanding as a schema of the divine understanding. We may infer from this that Maimon's monism, like the analysis of the concepts of affection and consciousness, is not meant to be read as a claim that the objective world is human, but that in the human mind there is a divine power. This is no diminution of the world, but rather a magnification of human beings, and, accordingly, there is no human-God dualism.[7]

The claim that the divine understanding is the source of the manifold of intuition translates into the claim that humans analyze intuitions into their conceptual components. The temporal order requires that we regard understanding as a result of intuition, but it is only the logical order, that is, the priority of the objective over the picture of the objective, that drives Maimon to hold that the intuition is an expression of the understanding: "I, on the other hand, regard them [the objects of metaphysics] as actual objects, which, although they are just ideas, can be thought of as determined by means of the intuitions which derive from them. By establishing the foundations of intuitions, we can determine new relations amongst them, thereby treating metaphysics as a science" (*Versuch* 196).

We thus reach the fourth point listed in Maimon's summary of the main points of his work in the introduction to the *Versuch*. The possibility of understanding finite human consciousness as one aspect of an understanding that creates the manifold of intuition from within itself is what enables, or rather forces, Maimon to avoid transcendental deduction.

One could explain Maimon's avoidance of the Kantian attempt to connect the possibility of self-consciousness with the lawful nature of the world by a different route. For Maimon, the magic, or the temptation, lies in a more thought-provoking metaphysical thesis, namely, the classical idea of the unity of knower-known-knowledge. Immediately after presenting his position, Maimon takes a hesitant step in this direction: "We have here before us (if I may use such a phrase) a triple-unity: God, the World, and the human mind." He goes on to explain: "In other words, if, by the world, we understand only the world of the understanding, that is, the sum of all possible objects which can be created for all the possible relations which the understanding can cognize; and if, by the mind, we understand an understanding (a cognitive faculty) which refers to these objects in such a way as it can think of all these objects; and if by God we understand an understanding which actually thinks all these relations (otherwise I do not know what I should think of as the most real entity), then all these things are one and the same" (*Versuch* 207).

This position is consistent with the essence of Maimon's system, rejecting the Kantian ideas of God, the World, and the Self, posing instead a single idea—namely, the Infinite Understanding.[8] It is also consistent with Maimon's interpretation of the concept of idea, that although it involves an inner contradiction, it is crucial for the complete analysis of human cognition. As far as the deduction is concerned, it would seem that Kant conjoined the human mind—or, at the very least, the representation of the "I think"—with the lawfulness of the world, thus leaving God out of the picture. Maimon's dissatisfaction concerns this point, as it motivated him to search after such connections between the human mind and the world that do not render God redundant.

KANT DESCRIBED the difference between his system and that of Maimon in terms of the notion of intuitive understanding. Maimon viewed this understanding as an idea, while Kant thought that this understanding is comprehensible to us only by negation. But this way to formulate the difference is not telling, as it represents Maimon's philosophy as one that has not fully grasped the transcendental shift. This is perhaps the source of Allison's evaluation of Maimon's system as theocentric (1983, 60). To understand better the power of Maimon's system we should rephrase the dispute between him and Kant in terms of the notions of consciousness and outside consciousness. Thus, the difference between Kant and Maimon concerns the coherence of the phrase "outside consciousness": Kant invokes it, while Maimon combines rationalism and idealism and offers a substitute for it. The ground for this change in terminology is obvious: An intuitive understanding is by definition a faculty that produces the multiplicity of phenomena, and in the course of doing so cognizes it according to its forms; as such it does not require any material from the outside. The merits of this formulation are clear as it captures in a concise way the main philosophical moves Maimon made in his intensive dialogue with Kant's first *Critique*.

7

Substance and Causality

MAIMON OPERATES his polarity on Kant's categories, claiming that what was already discovered for space and time in the case of the straight line applies here, too. The relation between the substance and its accident is the picture of a relation of determinability, and the relation between cause and effect is a picture of a relation between differentials of understanding. However, unlike the case of the straight line, but in accordance with the view on space, we have no certainty that there is any objective relation between what we see as substances and their accidents or causes and effects. It all may be a product of imagination.

Subject, Predicate, and the Law of Determinability

One of the issues that recurs throughout Maimon's writings, especially in his later work, is the relation of determinability amongst concepts. This relation, which we have already encountered in explicating the quid juris question, is used by Maimon to define the relation between subject and predicate in such a way that it will fit in with his strict concept of objectivity.[1] Maimon's main focus in this context is on the distinction between two kinds of syntheses of concepts. One is the random conjunction of one concept with another, as in the case of *black triangle.* In the second type of synthesis, one concept is the determination of the other, as in the synthesis of *straight line.* In this case, Maimon calls the concept *line* "the subject of the synthesis" and the concept *straight* "the predicate of the synthesis." "If a synthesis is of this kind, whereby one of its parts can be thought of without reference to the other, that is, it can be thought of either in and of itself or in another synthesis, whereas the second part can be thought of only in relation to the first, then the first part is called the subject of the synthesis and the second part is called the predicate of the synthesis" (*Versuch* 84).

Maimon thus renews the relation between genus and species, but the idealist-transcendental context imbues this distinction with a special significance. This is clear from his definition of what is involved in the concept of determination, by appealing to the concept of thought.[2] On this point, Rotenstreich (1939) deserves a special mention, for he discusses the role of this principle in Maimonic thought and the fact that it enables one to pinpoint Maimon's philosophy between the thought of Kant and that of Hegel.

The relation of determinability is analogous to the relation of correlates, which Maimon employed in his definition of pure cognition. But, unlike the relation of correlates, in which each concept is both the subject and the predicate of the other, the relation of determinability is characterized as directional and asymmetrical, owing to the difference between the subject and the predicate. In the case of the conjunction *straight line,* this asymmetry is reflected in the fact that one can understand the idea of a line by means of reference to the conjunction circular line, but one cannot understand straight without referring to the relation between this concept and the concept of *line.*

The relation of determinability, just like the relation of correlates, has validity for all thinking beings. If, for one thinking being, a given concept is determinable with regard to another concept, then these two concepts hold this relation for any thinking being. This is what motivates Maimon to define the relation in terms of the possibility of thinking. In the case of syntheses that are not in accordance with the law of determinability, or concepts that are not connected by means of the relation of correlates, the grounds for the ensuing unity are unclear, for they can be thought of in a different way.

The relation of determinability, then, is an objective relation. Note, however, that objectivity refers us not to the concept of object but rather to the concept of thought. In the relation of determinability, the subject is not the thing that carries the accident, but a concept that it can be combined with another concept. The relation of determinability is most significantly employed in the course of the explanation of the substance–accident relation: "The concepts of subject and predicate, when employed for empirical objects, grant us the concepts of substance and accident, that is, if an empirical object (an intuition) can be thought of in various different syntheses (and as it is an empirical object, we cannot be certain of the possibility of thinking of it in different syntheses unless it is actually given in these syntheses), it is called an object, whereas the different determinations of this thing, with which it is thought of in synthesis, are called accidents of it" (*Versuch* 95).

The substance-accident relation, which we encounter empirically, is a picture of the conceptual and objective relation of determinability. However, the skeptical dimension in Maimon's thought does not allow him to apply this relation apodictically to empirical objects. He preserves the possibility of the substance-accident relation, but he is not certain at all whether objects investigated by the empirical sciences *actually* observe this relation:

> Gold, for example, is a concrete synthesis of the color gold, a particular weight, density, etc. There is no conceptual synthesis involved in this at all, as these properties do not stand in the relation of subject to predicate (determinant and its determinability), for one can think of one without the other. Rather, they are thought of together only because the one goes hand in hand with the other in time and space. I readily admit that it is necessary for a synthesis of the faculty of imagination to have an inner reason, that is, that a mind which knows the inner essence of gold must create for itself a concept about it such that these properties necessarily follow from its essence. However, from our point of view, this synthesis will always be merely a synthesis of the faculty of imagination. (*Versuch* 391)

Maimon regards gold as a synthesis of properties, produced by the faculty of the imagination. The synthesis involved is not arbitrary, but actual, in the sense that it is imposed on us by our imagination: "The merely possible, as opposed to the actual, is what has been invented, that is, a synthesis which we did not feel but which was executed in a completely arbitrary way, for example, the color green, a particular weight, etc. The possible is distinct from the actual not in terms of its type, but in terms of degree, in other words, from the point of view of the less common proximity in space and time, or from the point of view of the weaker validity of the images themselves" (391). The synthesis is not an arbitrary one, but we cannot be certain that it is intelligible, for it is possible, albeit in a problematic sense, that there are other subjects who analyze reality differently. The object which we perceive in space and time is a conglomeration of properties that owes its reality to our faculty of imagination.

We will never be able to turn what we think about gold into an objective definition of gold. We cannot apply the points made about the relation between a priori cognition and pure cognition to the empirical realm. Whereas the a priori properties of the straight line were sufficient for us to reach a definition of the straight line as a property of being the shortest distance, we cannot, by the same procedure, make the phenomenological properties of gold into a definition—certainly not an apodictic one. The objects of the empirical world do not provide Maimon with the qualities that he managed to find in the case of mathematical configurations.

This critique of Kant reveals some of the rationalist-skeptical aspects of Maimon's thought, and it is thus easy to overlook the idealistic level. The latter aspect, as we shall see, was not emphasized by Maimon's commentators and deserves some attention. Maimon's idealism is reflected in his account of the concept of object in two levels: the first is the fact that there are no objects outside consciousness, and the second is the restriction of the cognized object to the finite potential of human consciousness.

Before expressing his views on gold, Maimon explains: "An actual entity (Das Wirkliche) is not, as some philosophers believe, something determined in all its aspects (ens omni modo determinatum), for were I to admit that everything actual is a completely determined entity, this would not entail the opposite, namely, that every completely determined

sense, Maimon introduces an antirealistic position into Aristotelian logic. His renunciation of this principle is not motivated by the fact that the concepts are vague, or that certain concepts (such as green and prime number) do not relate to each other, but follows as a logical conclusion from the analysis of the finite nature of human cognition, which is incapable of encompassing an infinite number of concepts. But Maimon goes much further in interpreting the finite potential of human cognition, insisting that even the clear-cut rules of inference need to be corrected. "The connection between thought and the unity of consciousness is such that without the unity of consciousness there is no thought. The same applies to the self (the identity of consciousness at different times). The self must, naturally, be identical with itself at different times, for otherwise thought would be impossible . . . And the same is true if, for instance, I thought that *A* is *B*, whereas another thinking being would think the sentence *B* is *C*—it will never follow as a conclusion that *A* is *C*" (*Versuch* 156).

Thought is a synthesis of components that are, in the final analysis, possible determinations of the subject. But in order for these components to form a thought, they must be attributed to a single subject. Here Maimon uses the notion of the transcendental deduction, but not to ensure the possibility of experience, rather to guarantee the possibility of thought. To this Maimon adds his analysis of objectivity. The conclusion of an inference does not hold by virtue of the premises alone. The objectivity of the inference is such that whoever thinks of the premises and the conclusion cannot reject the conclusion. This addition attests to the radical nature of Maimon's idealism and makes it prone to psychologism.

Causality and the Differentials

The relation between determinability and the category of substance is similar to the relation between the differentials and the law of causality. The differentials and the relations amongst them are the necessary intellectual relations on which the relation of causality must be founded for it not to be a mere association created by the faculty of imagination, as in Hume's theory, or a connection of syntheses of the faculty of imagination with the pure category, as in Kant.

Maimon embarks on a far-reaching venture here, endeavoring to connect calculus with his theory of sensation and intuition, thereby treating it as a conceptual entity on which the intuitive relation is founded. As a result, he ascribes certain properties to differentials in such a way that it is difficult to understand how they fit in with each other. In fact, Maimon himself acknowledges the obtuse nature of his theory of the differentials and abandons it in his later work. The difficulties involved in this theory increase in light of the accounts of the concepts of limit and derivative provided by late nineteenth-century mathematicians. Even Robinson's nonstandard analysis, which designates differentials as mathematical objects, cannot be of any assistance in better defining Maimon's position. For this reason I have refrained from assigning the differentials a central place.

Maimon draws a distinction between two concepts of infinitely small. The first is the position that quantity approaches but can never reach (*Versuch* 352). The second is *any* position that a quantity can reach. He demonstrates the first type by means of the angle between two parallel lines. We can apprehend this angle as the limit of angles approaching zero. In this case, one can talk of an infinitely small angle, but when we come to the limit, we do not actually have an angle. We can talk of a zero cosine, but "only in a symbolic sense" (352). The second type is the infinitely small, or differential, which is metaphysically oriented. This differential is not the absence of quantity, but "any position which it can reach indiscriminately, that is, an undetermined position which can be determined. If, then, I say: $dx:dy = a:b$, this does not mean an x abstracted of all size relating to a y abstracted of all size, etc., for nothing cannot relate to a nothing in a relation of size. What this means is as follows: however large or small we assume x to be, from the equation between the two sizes it always follows that $x:y$, etc. Hence I assume here that x is smaller than any given size, and from this it necessarily follows that $dx:dy$ and so forth" (*Versuch* 352).

In order to understand what Maimon is doing here, we have to imagine that x and y are two sides of a triangle. We can sometimes prove that these sides stand in a certain relation to one another. In such a description, we are not arguing for the existence of a connection between four numbers that correspond to four particular sides, but for the existence of a general relation. We thus determine some general connections between the sides of a right-angled triangle while determining certain relations of proportion. This general law is expanded at the second stage, as evidenced by the conclusion of the above excerpt, to cover the differentials, which are perceived as sizes smaller than any given size. Here the differentials represent something that can be determined, i.e., determinables.

Already in the metaphysical sense there are two interrelated concepts of differentials. One is the well-known idea that the differentials are magnitudes smaller than all positive numbers but not equivalent to zero. The second idea, which is unique to Maimon and is as interesting as it is obscure, makes a connection between the differentials and the notion of a state that has not yet been subject to determination. In the second sense, the differentials are like genera, which can also be determined. But unlike genera, which are usually forms that can be determined by a particular concept, the differentials are a kind of polymorphous form, which can be determined in a great number of places. Instead of talking here of genus and of being determined by type, we are now talking of an indeterminate relation, which can become determinate by means of a number of components or magnitudes. In this sense, the differentials are connected to the relation of determinability; in effect, they extend the relation of determinability to relations in general.

Accordingly, a differential equation is a conceptual law on which observable, quantified relations are based. Maimon explains how we get to this law amongst differentials by means of expanding the geometric law to differentials; but from the point of view of logical sequence, he suggests a complete turnabout: the differential relation lies at the base of the relation revealed to intuition.

The metaphysical infinitely small, according to Maimon, is real since one can observe the quality in and of itself while abstracting it from all quantity. In thus employing the differentials, Maimon does not regard them as small units but as abstractions from quantity. He believes that with the aid of this concept, one can solve the quid juris question: "Pure intellectual concepts or categories never refer directly to intuitions, but merely to their elements, which are ideas of the understanding . . . and by means of these foundations they refer to the actual intuitions. Just as in higher mathematics we deduce the relations of various magnitudes from the differentials of these magnitudes themselves, so does the understanding deduce (albeit gropingly) from the real relations of the differentials of various magnitudes the real relations of these qualities themselves" (*Versuch* 356).

Maimon immediately goes on to illustrate this point by means of the proposition "fire melts wax." This proposition does not apply to fire and wax as objects of intuition, but to their elements. This description is consistent with our premise, and, indeed, Maimon clarifies this point in referring to an example from arithmetic. The differentials, like numbers, are intellectual determinations, and the understanding is thus entitled to keep on subordinating them to pure concepts.

It is quite difficult to comprehend Maimon's leap from quantitative relations granted by differential equations to intuitions or objects like fire and wax. Maimon, in effect, is adapting a relation typical of geometrical configurations to the empirical realm. One may be able to comprehend the idea of a right-angled triangle with sides of differential lengths, but it is quite difficult to comprehend the idea of the differential of fire. Furthermore, it is not quite clear how one is to understand the connection between intuitions and integrals. If we think of this connection as simply an example of the fact that the relation between differentials is transposed to the relations between various different entities, then it is not entirely clear which element of the example corresponds to the relation between the differentials. On the other hand, if we take intuition to be the integral of the differentials, we will then have to consider the differential as a small magnitude—somewhat like Leibniz's "small sensations," the conjunction of which gives us consciousness. But in this case we would be dealing with the mathematical, not the metaphysical differential.

Nonetheless, Maimon's intention in employing the differentials is rather obvious. He argues that a causal connection, like the connection between lighting a fire and melting wax, has to be based on a universal law of nature. Certain causal relations must follow deductively from natural laws and from the definitions of the objects that feature in relations of cause and effect. Modern science has made considerable headway in this field. We must deduce causal relations in the same way that in modern science we deduce physical and chemical phenomena from the atomic theory. According to Kant, causality consists in attaching a category to a connection of the faculty of imagination. This analysis, says Maimon, is insufficient. Maimon seeks to attain something similar to what was achieved by realistic-materialistic theories in physics within an idealistic-rationalistic framework.

Kant, too, held that the concept of causality derives from logic and that, as such, one could think of using it in regions unlike our sensuous forms. However, such uses have "merely a logical significance, attached to the unity of representations alone. No object is given them, and therefore they have no meaning sufficient to afford us a conception of an object." He goes on to say: "The categories, without schemata, are merely functions of the understanding for the production of conceptions, but do not represent any object. This significance they derive from sensibility, which at the same time realizes the understanding and restricts it" (*Critique* A-147, B-186). Kant believed that restricting causality to the conditions of sensibility would necessarily limit the validity of certain causal judgments to beings sharing this form of sensibility. According to him, "although the schema of sensibility are the sole agents in realizing the categories, they do, nevertheless, also restrict them" (A-147, B-186). But Maimon, who demonstrated that the form of sensibility has an objective basis, is not bound by this restriction, which is imposed on Kant. Kant proceeds from the category that, for him, is a form of sensibility, whereas Maimon prefers to start from that which is apparently embedded in the form of sensibility, and to search for its objective basis.[4]

MAIMON OFFERS different summations of his system, and toward the end of his tract he also makes a few comparisons: "There is a point where materialism, idealism, Leibnizianism, Spinozism and even Theism and Atheism can coincide (if only these gentlemen understood themselves and refrained from viciously agitating the masses against their rivals)" (*Versuch* 208).

The metaphilosophy presupposed by this unified system is that grains of truth can be extracted from the various philosophies that cannot but be mutually compatible. These grains of truth, however, do not leap to the eye. A great effort is required to excavate them. There is a common point, which is not the mere cross section of these different philosophies. Maimon conceives it as an imaginary burning point, namely the imaginary meeting point of light beams. If we regard the basic intuitions of each system as rays, then by continuing the line of each ray we can find the imaginary focus from which the rays spread out.

What is not mentioned, however, is quite conspicuous: Hume's skepticism. In other places Maimon declares that all he tried to do was create a synthesis between Leibniz, Spinoza, Hume, and, of course, Kant. If we adhere to the rational-idealism, then this rationalism is the imaginary stake where all nonskeptical systems are edified. We can now go through each philosophical system and see how it is treated in the rationalist idealism.

The materialist is justified when he argues against the idealist that there is something out there beyond the phenomenal, but he is mistaken to think it is independent in all ways from the structure of consciousness. If we take into account Maimon's reconstruction of the outside in terms of the ignorance of the conceptual component that lies at the foundation of the intuition, we find that the materialist is not aware of this and makes an unnecessary leap in assuming the existence of material objects.

The idealist is right in rejecting the notion of material objects, which exist in space and time independently of our consciousness. Leibniz is correct when he views intuition as mixed cognition, and Maimon helps him with the idea of Maimonic reduction to clarify his position. Leibniz is wrong when he thinks God the absolute reality and man only one of the monads. Spinoza is correct in the parallelism of the attribute of extension and the attribute of thought, but one should replace *extension* with *intuition.* He is also mistaken in thinking God is the substance. Similarly, the theist is right to think the world is in God and that without God there is no world, but he is mistaken to think of God as an object. In the same manner, the atheist is correct in his condemnation of the idea of a god that exists independently of man, but he is mistaken when he continues from this point to state that this idea is unnecessary; infinite understanding is the limit of man, and only by assuming him can we overcome the dualism we find in scientific consciousness.

Maimon also explains how we can incorporate Kant into this rationalist idealism. From the viewpoint of the finite consciousness, Kant is correct in distinguishing between the species of intuition and that of cognition. However, from the viewpoint of the infinite understanding, or when we take into account the possibility of development in science, the partition between judgment on the basis of concepts and judgment based on intuition is not absolute. Nevertheless, it is clear now that Hume is not a party to the Maimonian synthesis. In the terms of Maimon's metaphor, the ray of light, Hume's insight is not projected from the imaginary stake.

Maimon, then, doesn't have a theory that can integrate skepticism and rationalism. Beiser's analysis of Maimon as vacillating between skepticism and rationalism stems from the identification of the skeptical aspect with the assertion that there is a gap between the cognitive and the sensuous. In this context, Beiser's identification of the choice between rationalism and skepticism with the question of the existence of the infinite understanding is not accurate. If this were the question, then the image of the infinite understanding as fiction, or as a regulatory idea, would be the way out (what Beiser calls the middle path). But what we called here rationalist idealism expresses the isolated rationalistic aspect of Maimon. The skeptic doesn't point out the gap between the intellectual and the intuition but asks whether there is any intellectual component at all in our intuition. And that indeed is the reason why Humean skepticism cannot be seen as another component of rationalist idealism.

Nevertheless, the rationalistic option in the background of Maimon's work grants this skepticism a special status. Hume's skepticism regarding the concepts of causality and object is not insurmountable. One can take it as a starting point for an account of the concepts of object and cause, as is the normal practice in naturalist empiricism. We begin with the belief that the causal relation is an objective relation inherent in things. This belief is undermined, and, eventually, we find ourselves left with skepticism. However, this skepticism itself turns, at a later stage, into the analysis of the concepts of causality and object: causality consists of associative connections amongst images, and the object is a collection of properties closely connected in our imagination. In taking the step from skepticism to analysis, the naturalist empiricist is assisted by the attack on the very rationalist requirement to anchor lawfulness in a normative foundation.

This process of thinking is recurrent in many philosophical pursuits. We think of concept *A* and associate it with characterization *B*, which we are unable to substantiate, so we move to an analysis of *A*, which excludes *B*. For example, we look for a criterion for determining absolute movement and eventually we are satisfied with relative movement.[5] In contrast to Hume, Maimon leaves us with skepticism that cannot be removed by resorting to descriptions: the rationalist element enables us to inquire into the concept of an object, and objective interrelations, which go beyond associative connections. Thus, while Hume's skepticism, once it moves toward descriptions, leads us to reject the claim that we cannot know anything about causality, Maimon concludes that we are restricted by knowing what it is that we lack.

In this way Humean skepticism becomes insurmountable. This is the core of Maimon's unease, which is a consequence of the fact that our condition is now defined as one of ignorance. We lack a picture, or a clear horizon, against which to assess the possibilities inherent in our search. We cannot embark on an optimistic rationalist voyage in the course of which the rational essence of the universe will be increasingly revealed to us.[6] Yet we cannot allow our skepticism to nourish the claim that the very search for the rational foundation is meaningless. Like Socrates, Maimon aspires to lead us into the position in which we will have reason to believe that we do not know. This lack of knowledge may never become clear knowledge. But it is sufficient to establish our tendency to make assumptions, to explore, to try and approach knowledge, and to be constantly striving toward the rational basis of our world, however elusive it seems to be.

We now see how Maimon demonstrates the possibility of metaphysics and at the same time, and with irremovable tension, maintains the coherence of Hume's skepticism. Skepticism exists and doesn't disappear just because a metaphysical answer is possible. Be that as it may, we gain the assertion that metaphysics is not the impossibility Kant thought it is.

Kant is the philosopher trying to defend the common people against scoundrels. This is his guiding metaphor. Kant is also the censor, the critical thinker who puts reason on trial. Maimon is not a partner to the endeavor to make metaphysics into a science. He dedicated his life to metaphysics but realized that it would never reach a peaceful shore. Unlike Kant, who regarded logic as a theory to which there was nothing to add, and Newtonian science as the solid foundation of human science, Maimon regarded logic as infinitely expandable and science as far from immune to upheavals.

Maimon's starting point is different from Kant's. For Kant, the metaphysical disputes are an indication of the helplessness of human understanding in confronting the problems it faces. Maimon, on the other hand, views the positive side of these disputes. For him, the effort involved in them is an immanent component of human destiny and as such it cannot be futile. At the end of his autobiography, Maimon described the entire history of philosophy as the attempt of dancers to dance with the maiden, who represents metaphysics. Kant could not have even envisioned such a metaphor: for him, the speculative endeavor has to serve a definite, practical purpose, for otherwise it would not be useless.

Maimon's methodological course supports this conception. Naturally, his rationalist leanings led him to present progress as an immanent element of cognition. The objective world is the conclusion of the search for the foundations of the sensible world. But we would be mistaken in thinking that we can be assured of the success of this search. It may well be that the whole idea of search and progress is nothing but an illusion. We cannot find rest in the very wagon sent to search for the grounds of the sensible world.

Epilogue

Closing the Circle
Maimon and the Kabbala

MAIMON'S SPIRITUAL journey, according to his own testimony, began with Maimonides and the kabbalistic *Heshek Shelomo*, Maimon's earliest work. It proceeded under the inspiriting influence of Hume, Spinoza, and Leibniz and eventually took up the challenge posed by Kant's philosophy. By the end of this journey, Maimon was no closer to making the transition from the sensuous to the intelligible world than he was when he set off, though in the course of time he did manage to come up with an original system. Yet, did his philosophical pursuit take the shape of a linear progress, as is implicit in his autobiography? If so, this would suggest that his involvement with the kabbala was only a temporary, though significant, stage in his life, "a monument celebrating the aspiration of the human spirit to perfection" (*Autobiography* 1:143).

In his autobiography, Maimon devotes a considerable number of pages to the kabbala. The long appendix he attached to this work consists of two parts. The first presents his views on Jewish esoterics; the second serves as a general exposition of Maimonides's thought. Even though Maimon favored Maimonides's rationalism, traces of Jewish mysticism can be detected even in his later writing of the *Essay*. Generally speaking,

Maimon reads the Kantian problematics in light of medieval thought. Conversely, in his commentary to *Givat hamoreh,* he reads Maimonides in view of the Copernican revolution. Influences of this sort are of interest here only insofar as they are helpful in elucidating those kabbalistic residues that linger on in Maimon's later thought.

One way of viewing the kabbalistic remnants in Maimon's philosophy merely touches the surface of his thought. For instance, he conveys his mockery of the Kantian order by establishing a correspondence between the twelve Kantian categories and the number of the Hebrew tribes. He thus channels his kabbalistic instinct to produce a subtle form of irony. Undoubtedly, for anyone attuned to the spirit of the Jewish Midrash and kabbala, an identical number of things that are subsumed under two separate concepts is a clear indication of the existence of a deep connection between them. In fact, in *Heshek Shelomo* Maimon himself drew a parallel between the ten *sefirot* of the kabbala and the ten categories established by Aristotle. In thus demonstrating the arbitrariness of the Kantian architectonics, Maimon projects his own tendency, which he eventually abandoned in the course of his pursuit of the truth.

At a deeper level, traces of Maimon's interest in the kabbala can be detected by the way he channels kabbalistic structures into his later philosophical concerns. The first parallel concerns the human ability to create real objects in the course of making judgments. Such a process of producing real thought *(reale Denken)* occurs when we form syntheses according to the law of determinability. For example, our judgment that it is possible to determine the concept of line by referring to the concept of straight gives rise to a real object: the straight line. The same applies to creating the objects of mathematics on the basis of mathematical relations. As a matter of fact, according to Maimon, the entire realm of pure cognition and our ability to actually create objects in the mind (and not merely to represent the objects of our perceptions) attest to the creative power of human understanding.

Maimon does not hide the theological sources of his thought. He makes it clear that as far as mathematics is concerned, we resemble God, and he elaborates on this point: "The creation *ex nihilo* is not entirely foreign to our way of thinking. We derive this concept from within ourselves. For through understanding, the human mind can cognize objects *a priori* and by means of the faculty of pure imagination it can provide them with *a priori* material or shape them as objects" (*Philosophisches Wörterbuch* 1:246).[1] Maimon's insistence that the ability to create is a human property is reminiscent of the kabbalistic notion of the operation of theurgic powers, which accounts for human intervention in the higher spheres. The book of the Zohar describes how the heavens are created through the prayers of the zaddik, and similar descriptions are recurrent in various kabbalistic writings. All in all, by attributing to the zaddik the powers of cosmic restoration the kabbala exalts human beings to an extent unmatched in Leibniz's and Spinoza's works.

Dogmatic rationalism viewed God's infinitude as an indication of the finitude of man, who is but one of the items of Creation. This human creature, whether as a Spinozian *modus* or as a Leibnizian *monad*, is capable of inquiring into the wisdom that created the world but is incapable of being a creator in its own right. Conversely, according to the kabbalistic outlook of the world, God's infinitude does not entail human passivity; rather, it allows humans to be God's partners in the creation of the world.

By ascribing to humans the ability to create objects ex nihilo, Maimon establishes an interesting symbiosis between the exaltation of humans in Kant's Copernican revolution and a similar exaltation of the zaddik in the kabbala. Maimon's logical idealism results, then, from the combination of two developments. The first lies in his acceptance of Berkeley's critique of the idea that matter exists outside consciousness, and in his response to Humean skepticism and the Copernican revolution; the second consists in exalting human beings on the grounds that they were created in the image and likeness of God. In this respect, the encounter with Kant finds Maimon well prepared. Kant should have been surprised by Maimon's determination to embrace the core of the Copernican revolution and to interpret it in a radical way. It is as if Kant says: it is true that I have exalted man, but one must acknowledge that human thought cannot create objects from within itself; this indeed was the grounds for his characterization of Maimon's philosophy as embracing intellectual intuition. In view of this, one can appreciate the difference between Maimon's thought and that of his contemporaries: while Kant's contemporaries were afraid he reinstated Berkeley's idealism, Maimon was ripe for contact with the Copernican revolution and was ready to engage in a radical extension of it.

What made this synthesis possible was the place Maimon assigned to mathematics in his thought. According to Maimon, mathematics deals with objects, and not with abstractions, as was the view shared by the medieval thinkers, and by Leibniz, Spinoza, Wolff, and Kant as well. While Maimon was grappling with Kant's thought, trying to salvage the divine nature of human reason, he realized that human ability to create objects does not detract from the objectivity of the human mind. In making this point he was assisted by his profound understanding of mathematics: the inherent properties of the contents of mathematical knowledge do not entail recourse to matter.[2]

It seems that Maimon's interest in the kabbala has played an important role in the unfolding of his program of the *Versuch*. This role can easily be detected if we compare Maimon to Frege, who shares with Maimon some major assumptions (about the connection between reason and objectivity, on the one hand, and intuition and subjectivity, on the other) and an anti-Kantian orientation. Frege adopts the claim that numbers and reason are closely akin but criticizes the formalist, who sees himself as God on account of his ability to create objects by laying down postulates (1950, 119). In Frege's view, the mathematician is not an inventor but rather an explorer. Maimon subscribes to the rationalist position about the role played by intuition in forming the concept of number, while still defending an idealistic position. Had he not believed in the active participation of humans in the pure world, he would have tended to identify idealism with psychologism, as did Frege.

The connection between Maimon and the kabbala should not be overestimated. One must concede that in exalting human beings to the extent that God is reduced to a mere idea, Maimon goes against the traditional spirit of the kabbala. According to the kabbala, man is created in the image of God, while God is the very foundation of the world. Nevertheless, at times the theurgic orientation in the kabbala reaches such heights that the Godhead, or at least some of its aspects, is regarded as being intensified, or even actually created, through prayer. Idel (1988, 201–3) distinguishes between a "status quo theurgy" and a "creative theurgy." In the former, the zaddik, or the strict observance of the commandments, preserves the balance between the upper and the lower worlds, whereas in the latter the action of the zaddik actually creates the pure world. Creative theurgy is discussed in the Zoharic explanation of "va'asitem otam": "He who fulfills the commandments of the Torah and walks in God's paths creates Him, as it were, in the world above. The Holy One, blessed be He, says: *It is as if he created Me (ke'ilu a'saani).* Therefore, it is written 'And you shall do them (*va'asitem otam*)' (Lev. 26:3). It is not written *otam,* but *atem* (you)—literally" (Tishbi 1961, 1160). This radical exaltation of man and the perception of human intellect as the schema of the divine creative intellect suggest the affinity between Maimon's notion of the infinite understanding and the outlook of the kabbala. According to the interpretation suggested above, Maimon views the infinite understanding not as an object but as a limiting concept that demarcates the boundaries of human development. Hence Maimon reads "*ke'ilu asa'ani*" literally.[3]

A second close proximity between Maimon's later ideas and the kabbala is to be found in the very idea of the Maimonic reduction. In the kabbalistic doctrine of the Ari (R. Isaac Luria) with which Maimon was familiar, the Restoration (*Tikkun*) is a process of purification that takes place within the dimension of time and reinstates things to their original perfection. The Maimonic reduction, too, is a purification (of the intuition) that operates within the dimension of time, and it has a similar effect, as it uncovers the pure conceptual element—which is analogous to the kabbalistic notion of light. In exposing the foundation of intuition, Maimon raises the level of objectivity and makes it transparent to consciousness.

Lurianic mysticism places emphasis on the outflow from this world to the upper worlds. Rationalistic positions stress an opposite movement. The general direction in the Leibnizian or Spinozian positions is downward, which is tantamount to the descent from the pure to the impure, as discussed in medieval thought and later by Maimonides. The general course of the Maimonic reduction follows that of the kabbala: from the impure to the pure. In view of this, we can complete our account of Maimon's reasoning by pointing at a classic theosophical direction: if human activity consists in a continuous process of unraveling that gradually uncovers the pure element—the foundation of sensibility—this may well be an indication of an earlier process of encoding, whereby the pure was imprinted in intuition. For surely it is not possible to have decoding without prior encoding, repair without damage (or breaking), and redemption without exile!

The Maimonic reduction and the notion that human beings have a share in intuitive understanding are isomorphic to some well-known kabbalistic ideas. This goes beyond the fact that some kabbalistic structure has found its way into the Maimonic universe of discourse. In this work we have analyzed the quid juris question while placing emphasis on its mathematical context. Yet, we know that for Maimon this question was of a broader scope and greater interest. When he introduced this question, he explicitly identified it with the problem of explaining the possibility of the descent of understanding into this world. Even when Maimon read Spinoza, he felt the need to confirm that Spinoza's teaching was in line with the Lurianic doctrine of the *Zimzum,* the divine contraction. We can pinpoint the nature of the kabbalisitc influence exerted on Maimon if we define his positive, nonskeptic philosophical approach as rational idealism. The kabbala allowed for the idealistic dimension in Maimon's system. Maimon's thought moves back and forth from philosophy to theosophy and mystical theurgy. The implied notion of a linear progress of his spiritual development is misleading and has to do with the fact that Maimon addressed his work to the German public. The truth is that Maimon's thought is multilayered.[4]

Many modern readers compartmentalize philosophical texts according to the medieval thought, the kabbala, the philosophy of mathematics, Spinoza, the Copernican revolution, and other divisions. Maimon does not recognize such boundaries. Not only did he project his thoughts into the kabbala, as did the Aristotelian kabbalists (he abandoned these ideas when he became disenchanted with the kabbala, as testified in his autobiography), but he also borrows ideas deriving from his Jewish heritage. Such a bilateral flow of ideas builds interesting bridges between philosophies and ideas that belong to different universes of discourse. For Maimon, while thinkers such as Cordovero, Leibniz,[5] Spinoza, Hume, and Kant[6] authored distinct philosophical texts, widely ranging in their methodologies and interests, all of these texts together form part of one single text. For him a continuous reflection upon this all-encompassing narrative is the task of any serious thinker.[7]

Appendix

Maimon and Modern Structuralism in Mathematics

> If in the consideration of a simply infinite system *N*, set in order by a transformation *F*, we entirely neglect the special character of the elements; simply retaining their distinguishability and taking into account only the relations to one another in which they are placed by the order setting transformation *F*, then are these elements called natural numbers or ordinal numbers or simply numbers. . . . With reference to this freeing the elements from every other content (abstraction) we are justified in calling numbers a free creation of the human mind. (Dedekind 1888 [1963])

This stance is not a matter for historians only. Some eighty years after Dedekind, Benacerraf returns to this view: "A number draws its individuality from its place in the series of numbers, or from its relation to all other members of the series and the rules that govern this series when we perform an act of counting" (1965, 293).

At this stage we already have the following results: (1) Structuralism in mathematics begins with Maimon in the eighteenth century, and not with Dedekind as it is usually assumed (by Parsons and Resnik); (2) Maimon's position is surprisingly like Dedekind's. Indeed, there is a connection between what Dedekind said and Maimon's view that "the numbers one and two, in and of themselves (abstracted from their use) are a priori relations which determine the objects of arithmetic. They have no inner properties whatsoever apart from their relations" (*Logik* 1 n). Another similarity is found in Dedekind's view that we are of a divine kind and possess without any doubt creative power not only in the material realm (railroads, telegraph) but most especially in the mental.

Moreover, even the context is similar. Dedekind was also interested in diminishing the role of intuition so important to Kant, and he did this with his theory of numbers, which is Maimonic in character. (This may be grounds for speculating that Dedekind knew of Maimon's work.) The important question is what we can learn from Maimon's version of structuralism and why it is important for a contemporary philosopher of mathematics to read Maimon on modern structuralism.

One benefit is that we can see the progress modern philosophers of mathematics have made. Moreover, since Maimon's thought is separated from the evolution that led to modern structuralism in the philosophy of mathematics, his work affords us an external point of view of what we are doing today, and we can see more easily the assumptions and the axioms of our discourse.

We can also understand Maimon better, as there are new questions that suggest different readings of known passages from Maimon and new emphases on aspects of his system. The comparison also gives us an opportunity to understand the relation between an insight and its philosophical context. We have here a similar recurring insight but different philosophical backgrounds (Maimon's, Dedekind's, Benacerraf's, and Resnik's): Maimon's work refers to Kant's transcendental idealism, while Resnik, Shapiro, and modern structuralists, for example, look to Tarski, Quine, and Frege. How does the context contribute to the meaning and significance of an insight?

Following is a hypothetical dialogue between Maimon and modern structuralism that revolves around these two fundamental questions:

1. What is a structure?
2. What is the ontological status of structures?

Before turning to these questions we must recall distance between Maimon's structuralism and the modern version. Modern structuralism was created to solve the problems of (1) the different reductions possible of numbers to set theory: any set with a single member can represent 1; (2) constraints from epistemology of mathematics, namely how we can know mathematical truths; and (3) what is known in the literature as the Julius Caesar Problem. Problem 1 is due to Benacerraf, who showed in his critic of Frege's view that a number is not identical with the representing set and that "number words are not names of special non-numerical entities, like sets, tomatoes or Gila monsters" (1965, 292). What is important about numbers are the relations, and any object can play the role of the number 7. Structuralism develops this idea by claiming that for numbers it is the texture of the relations between them that matters.

To the question of how abstract mathematical structures can be known, the structuralist answer is roughly that we can have a visual realization of structures. What we perceive to be true for that structure applies to all isomorphic structures. The last problem is also far from Maimon's interest and lies in Frege's requirement that every definition of number afford us with a way to determine that Julius Caesar is different from the number 7.

Maimon used his insight on the nature of numbers to develop a rationalistic-idealistic account of mathematics. Thus, as an idealist, epistemology was important to him. The modern structuralist takes it for granted that empirical knowledge is unproblematic. The reduction of truth of reason to empirical-like perception would be insufficient for him. In fact, he would like to emphasize the opposite direction, the reduction of what rests on intuitions to what is constructed by reason. Indeed, for Maimon all attempts to supply a theory of knowledge based on causation between the subject and empirical objects are dogmatic. We have a clear understanding of the properties and relations of numbers because we have posited them. They are transparent to us.

This, of course, is only the beginning of a possible dialogue between Maimon and modern philosophers. It is a dialogue that forces us to explain to ourselves what exactly is the problem of knowing truths about abstract objects, and to what extent we can view knowledge of truths about structures gained by visualizing a copy of the structure as empirical, etc. What is the nature of Maimon's positing that appeals to us? Is it the same notion we use today (for example, in the context of Quine's philosophy)? This dialogue is fruitful to both Maimon and contemporary philosophers of mathematics.

There exists in the literature two ways of explaining structures. The first consists of reflecting on a series of concrete objects and abstracting from all the inner properties of the objects, looking only at the structure. This is Dedekind's way of explaining structures. The result of this process is the creation of the system of natural numbers. This idea was criticized by Frege, who regarded it as a mentalistic view of numbers.

Resnik (1997) offers another way of understanding what a structure is by invoking an analogy with geometrical points. He states: "The underlying philosophical idea here is that in mathematics the primary subject-matter is not the individual mathematical objects but rather the structures in which they are arranged. The objects of mathematics, that is, the entities which our mathematical constants and quantifiers denote are themselves atoms, structureless, points or positions in structures. And as such they have no identity or distinguishing features outside a structure" (201).

Resnik then resorts to the notion of "position": "I take a pattern to consist of one or more objects, which I call positions that stand in various relationships. For generality I will allow that some of the relations may be monadic and that some of the positions may be 'distinguished'." (203). He explains this with the following analogy:

> A position is like a geometrical point. It has no distinguishing features other than those it has in virtue of being the particular position it is in the pattern to which it belongs. . . . Thus relative to the equilateral *ABC* the three points *A, B, C* can be differentiated, but when considered in isolation they are indistinguishable from each other and any other points . . .
>
> Here is how this geometric analogy applies to the natural number sequences (*N, S*). I take this to be a pattern with a single binary relation (successor) and the natural numbers to be positions. Viewed this way, there is no more puzzle to the natural numbers lacking identifying features beyond those definable in terms of this pattern than there is to the corresponding fact about the points in triangle *ABC*. (203)

It is not clear why we cannot compare points with other objects. No geometrical point is an elephant, and the first point from the following three is not identical to Julius Caesar. The analogy doesn't explain the impossibility of inquiring whether 7 is Julius Caesar, but it seems that it assumes that the question is meaningless. There is another problem: in the same way that we cannot compare 7 to Julius Caesar, we cannot compare Julius Caesar to 7. So perhaps Julius Caesar is a part of a structure and not the number 7.

I leave the discussion of modern structuralism at this point.[1] Here is what Maimon says about the nature of numbers: "The understanding, in my opinion, is merely the capacity to think, that is, the capacity to create pure concepts by making judgments. No objects are granted to it as material on which the mind has to operate . . . it is wrong to think that it is necessary for things (real objects) to be prior to the relations between them. The concepts of numbers are only relations, which do not presuppose any real objects, as these relations are the objects themselves" (*Versuch* 190). We have, then, a position about the relation between objects and their interrelations, what we would call an internal relations theory.

Here is a possible elaboration that suggests a different way to explain the insight that numbers are not usual objects but are constituted by the relations they satisfy. Maimon's notion of internal relations can be understood as an expansion of the determination relation. The determination relation holds when a concept is determined by another. An example of this is the conceptual compound *straight line.* This is quite different from composite noun phrases such as *nice donkey* or *school bus.* Maimon maintains that we can't think of the concept of straight without thinking of line. A determination is not an aggregate of concepts. When we have mutual determination, we have conceptual correlates, and we can't think of one part of the conceptual composition without thinking of the rest (see chapter 7).

The correlates manifest the notion of internal relations. The numbers 2 and 3 can't be thought of independently. To think of 3 as being a separate object, as Frege would want to think, is incorrect. Something in 3 suggests and points at 2. The relation between 2 and 3 is not on the same plane as that between 2 and Julius Caesar. Frege doesn't explain this affinity between the numbers. On the contrary, he would like to cancel it.

We can say this in an offhand way (that does not discriminate between objects and a proposition): 3 implies 2 and 5. This theory of Maimon is true not only about natural numbers but also about differential calculus. It is customary to say that only the ratio between differentials is meaningful, while in themselves differentials mean nothing.

Finally we come to the original question. A structure is a sequence of mutual determinations. This is not the same as other concepts, which are also located within a semantic network. If we take a concept—for example, "a mammal"—we will have a tree structure, connecting it with other concepts. The natural numbers are a system of mutual determinations. Rather than being in a tree structure, they are in a fully connected graph, where each element is connected to another.

I come now to the second question. The realists, or Platonists, maintain that these structures exist independently of the contents assigned to them. Eliminative structuralism holds that the structures are nothing beyond their embodiments. Resnik is a realist; Benacerraf is not. Where should we place Maimon in this debate? In so far as realism acknowledges the existence of entities independently of consciousness, Maimon was no realist. To him, the idea of existence outside consciousness is an absurdity. So, in this respect, Maimon would have rejected the Platonic independent reality of any structure. His reason for resorting to arithmetic as a battering ram against Kant was to show how it is possible to constitute objects by determining the relations among them. The same idea is operative in the definition of (nonmathematical) objects through combinations according to the law of determination. It would be difficult, however, to place Maimon amidst the eliminative structuralists. Maimon would deny that what *really* exists are concrete physical objects. Hume's skepticism regarding substance would prevent him from committing himself to the existence of such bodies.

Maimon discusses mathematical infinity for the purpose of discrediting Kant's antinomies. In Kant's response to Maimon's composition he urges Maimon to put his own system to the test by seeking a resolution of the antinomies within it. Kant is confident that Maimon's system of an intellectual intuition would collapse if it squarely confronts the antinomies. There can be no doubt, however, that Maimon studied Kant's remark very carefully, paying full attention to the crucial experiment that was to adjudicate between them, *before* stating his position in the *Ver-*

such. Indeed, Maimon's countermove is to extend Kant's *Ideas* and expand the antinomies far beyond their Kantian confines. Just before responding to the ideas in the *Versuch,* he actually refers to the end of Kant's recommendation and goes on to remark the following: "Whereas I greatly extend the domain of the ideas and the antinomies that they entail, by asserting that the latter can be found not only in metaphysics, but also in physics, and moreover even in the most rigorous science, namely, mathematics; thus the antinomies require a much more general resolution" (*Versuch* 227).

Maimon's diagnosis of what occasions the antinomies is as follows: "The complete series of numbers is not an object that can be given to my intuition but a mere idea, through which one regards the continual progression as if it were an object. . . . Reason here fastens upon a contradiction with itself, in that it sees something as an object, which, on its own terms, could never possibly be given as an object" (227). Such ideas are called by Maimon *ideas of the understanding,* which he contrasts with the ideas of reason. When we only have the rule of the progression, and we know that this rule cannot be actualized within intuition, as illustrated by the asymptote (*Versuch* 79), we have a pure concept, to which we know that no intuition corresponds.[2] We behave as if this rule picked out an object, yet we know that there is no appropriate object. For human reason, number requires an advance in time that can never be completed, while for an infinite intellect, relationships can be thought of at once, atemporally.[3]

Ideas of reason, by contrast, are such that even their rule is not given in a finite manner. A rule that involves conjoining an infinitude of concepts is not really thinkable, so it would count as an idea of reason; a given idea of understanding is a clear concept, except that we know it does not correspond to an intuition.[4]

To come back to our question on the ontological status of numbers, we will have to find out how the relation between an infinite intellect and our intellect is conceived. Such an examination is indispensable, for without it one might suspect that Maimon's solution to the problem of mathematical knowledge is merely nominal. And, indeed, the infinite sequence of numbers does not exist. They are constructions that man can create.

Now, for the needs of mathematical demonstrations, Maimon considered it unnecessary to presuppose the infinite totality of ideas entertained in thought. He accordingly proposed that we modify our definitions so as to meet these constraints. The set of all the lines (segments) constituting a circle is never *given* to us in any concrete form. The drawing of a circle by turning a segment round one of its end points gives us the consistency of the concept of a circle. It does not give us the multiplicity of lines comprising the circle. "The concrete completeness of the plurality cannot be given to intuition, because it is only possible to draw a finite number of segments equal to one another. The concrete completeness of the plurality, therefore, is not a concept of the understanding, but only an idea of the understanding, that can always be approached intuitively, by adding such lines; one by one to infinity" (*Versuch* 76). Consequently, Maimon makes the following proposal: "It is not necessary to include all the equal segments that share an end point in the definition of a circle; suffice it to require that whatever segment I draw from a certain point be equal to a specific segment, and thereby this concept will not be an idea" (*Versuch* 8).

Rather than a definition requiring the existence of all the lines, we will seek a definition suited to what we do in demonstrations, where we actually consider only a finite number of lines. There is no danger in such a move, for the following general reason: "if one realizes that mathematical concepts are not copies of something, such that we must check them against the original to ascertain their adequacy; that they are rather originals that the intellect creates from within, *a priori*, then their adequacy can be assessed only relatively, in respect of the conclusions that can be drawn from them" (*Versuch* 78).

The entire point of mathematical concepts inheres in the way they function in proofs and in the conclusions that can be drawn from them. When we examine our theorems, we do not need to presuppose for their proof that the circle is an idea of the understanding. By replacing the concept of circle by another concept, which is suited to our limitations yet yields the same conclusions, we avoid any substantial problem. This move of substituting one concept with another that yields the same conclusions will reoccur in the answer to the quid juris. But in the present context Maimon is utilizing it to achieve something else.

Maimon actually illustrates this by a triangle enclosed in a circle. The details of the theorem do not concern us here; what is important is that Maimon analyzes the proof of this theorem and shows through it that there is no need to presuppose all the radiuses as actually drawn, for the recourse to a finite number of points is sufficient. Thus, Maimon prefers to formulate the theorem requiring proof in a manner that does not refer to all the points of the circle but only to the few lines required for the proof. If, for example, we wish to prove the theorem that an internal angle spanning the diameter is always a right angle, we will reformulate the theorem so that it fits a proof demanding only a few lines rather than the entire continuity of the points of the circle. In modern parlance, Maimon is seeking the closest antirealistic substitute.[5]

We can now resume our initial question. Maimon rejects the idea that the structure of natural numbers exists out there. Natural numbers are the products of objective thought.[6] In all likelihood he would mobilize the same resources here, as in the case of the infinity of radiuses comprised in the concept of a circle. Today we know that such an adjustment would have led him to a different mathematics. Such a conclusion would not have deterred Maimon, who held that even logic should be adjusted to the constraints of strict idealism.

Maimon has no use for physical objects in order to *carry* structures in themselves. A number is a pure *object*, although its essence resides in its relation to other numbers. After suggesting his theory, Maimon tells us that he does not know what a number is: "We might pose the question, what a certain number is. It is not a posteriori object, since it's just a way of thinking of an object. It is not an a priori form, since it isn't a condition for an object. It is not a posteriori form, as this has no meaning, since forms can only be a priori. Therefore, what is it?" (*Versuch* 424). No one before Maimon took seriously the question of the nature of numbers, and despite the many ideas he had, which surpassed those of his times, he did not think of himself as having a sound answer to the nature of number.

Notes

Chapter One

1. Hintikka (1969, 1972) has taught us that the connection between intuition and sensibility is a *consequence* of Kant's philosophy, which does not follow from the definition of intuition. This entails that the more basic dualism is that of understanding and sensibility, but when one considers Kant's system as a whole and the relation between intuition and sensibility, the result is the intuition-concept dualism.

2. I do not believe that one can answer this question with an argument that starts out from clear, irrefutable premises and leads with logical necessity to the ideality conclusion. On the other hand, one cannot dismiss the arguments that Kant employs in order to reach this conclusion. In my view, Kant supports this conclusion with arguments incorporating postulates and a terminology he inherited from an earlier tradition. The following is one possible response to the above question, which will also introduce us to some of the key concepts of the Kantian system—concepts that shall feature prominently in the main part of this chapter.

3. An up-to-date discussion of this issue can be found in Buchdahl (1992, 135–65). Buchdahl constructs affection as a noncausal relation, and in this respect he follows Maimon.

4. Admittedly, there is something strange about explaining the concept of activity by an anthropomorphism of the understanding and the concept. It is doubtful this anthropomorphism can be eliminated by purges that would conserve the active aspect of concepts by saying, for example, that "the concept circle connects by its applying to many individuals," etc. Indeed, rendering the understanding as a homunculus is not a far-fetched metaphor. For Kant, and certainly within the heritage he received, the understanding and the "I" are inherently connected. Our putting together images in a certain way is understood as the necessity that the "I think" regulate all interweavings. The role of the "I think" consists in being a knitting needle, synthesizing representations into a world. Despite the vagueness of Kant's postulate of the activity of the mind, we cannot exclude risking the loss of a central layer in Kant's edifice.

5. Maimon, to the best of my knowledge, was unaware of the originality of Kant's conception of construction. This, however, does not affect Maimon's quid juris objection to Kant's position.

Chapter Two

1. I wish to thank Achim Engstler for helping me in the translation of this important passage.

2. Maimon modifies the concept substance in accordance with his strict idealist requirements, which ultimately permit the constitution of an object by joining together properties according to the law of determination. See my elaboration on Maimon's notion of substance in chapter 7.

3. I cannot even begin to discuss Maimon's unique conception of logic. A brief exposition is given in Beiser (1987, 309–17) and in Bergman (1967, 277–85).

4. This last level highlights an important aspect of all this taxonomy—namely, the idea of construction. Maimon's objection to any form of realism or extreme idealism does not allow him to regard the above categorization as that of theories with external objects of enquiry. Numbers are constructions, which do not rely on a priori cognition; geometry also contains creations, or at least reference to that which can be constructed; and practical mathematics deals with empirical construction—that is, technology.

5. My chief concern here is not whether such transitions from thought to the observable are troublesome or not, I only wish to separate them from the problem Maimon focuses on.

6. Some will object here to insisting that direction antecedes the concept of a straight line. This possibility does not necessarily present a severe obstacle to Maimon, since his main point is that the notion of straightness is independent of that of magnitude.

7. See Bergman (1967, 74). See also the comparison with Beiser in chapter 4.

8. Where, if at all, can we find this problem in post-Kantian philosophy? One may think of Frege's context principal and his theory of the unsaturatedeness of concepts. The connection between perception and the conceptual part of our knowledge is also relevant. Some aspects of Maimon's theory are also related to philosophy of mathematics.

9. Maimon did not question the success of the correlation between space and number, but rather the intelligibility of this correlation. From his perspective, a similar question would have arisen concerning the correlation between number and time, providing the former is constructed by the determination of the relations of the numbers in the mind, not by the intuition of time.

Chapter Three

1. I have been informed by several scholars that Hilbert was not a formalist at all, and that it was Brouwer who first called him a formalist.

2. Indeed, if Kant's insistence that the theorems of geometry are not analytic was sufficient for accrediting him with the anticipation of the idea of non-Euclidean geometries, then, by the same reasoning, one would be compelled to impute to him the weird notion of an arithmetic different from ours.

3. I do not propose to expound Maimon's notion of truth here, which deserves an independent study. I only wish to indicate its relevance to formalism as discussed above.

4. Inevitably, he must have recognized that one can consistently assume that the sum of the angles in a triangle is not 180 degrees. Otherwise it would not be consistent to assume that the outer angle is not equal to the sum of the two angles not adjacent to it.

5. We face the same vagueness when addressing the *truth* of the axioms. Hilbert contended that consistency sufficed for mathematical truth, but in the absence of an analysis of analytical truth, it appears as if he is equating mathematical truth—analytically—with consistency. One may wonder why he thinks it important to conserve the concept of truth at all, in addition to consistency. The way he deals with truth seems to contradict his showing interest in it.

6. One may be able to clarify this difference by means of the Fregean distinction between first-order predication, in which a property is ascribed to an object, and second-order predication, which applies to concepts of the first order. On the basis of this distinction, Hilbert's axioms are second-order predicates on the first-order concepts such as point and straight line. This implies that the Hilbertian transcription of geometry is essentially the following mechanical procedure: determine the axioms of geometry and replace every instance of a concept with a variable of the same type. In other words, if the concept is denoted by a one-place predicate, replace it with a variable or a schematic letter in which one-place predicates will be inserted, and so on. By this process, we get from propositions to second-order properties of the concepts that occur in the propositions. In this context, the distinction between Hilbert's formal study and correlative propositions becomes somewhat clearer. Correlative propositions are propositions whose truth is obvious and necessary. Second-order properties, which Hilbert presented as axioms, are not, by virtue of there being properties, true or false at all.

7. Compare with references in *Givat Hamoreh* (Maimon 1967, 27–29). The first is in Maimon's remarks on the introduction to Maimonides's *Guide,* and on Maimonides's commentary on Jacob's ladder, and the second is in his remarks to chapter 15, part 1 (48–52). In the first instance, Maimon provides a traditional analy-

sis of the ladder: climbing the ladder means ever-increasing abstraction and descending it designates ever-increasing determination. In this interpretation, Maimon distinguishes between the divine intellect, which is unlimited, and the ladder, which is limited on both extremes. Ascent and descent here connote the connecting of the truths. The second instance—that is, the second discussion of the ladder (chap. 15)—is already influenced by Kant; Maimon is trying there to explain his solution to the quid juris problem. What I have called here Maimon's ladder is closer to the second interpretation. So one can observe the Kantian impact between the two ladders, as it were.

8. According to the general concept of correspondence, which is simply the modern concept of a function, every phrase—be it "table" or "frog"—may correspond to "a straight line."

9. The question is typically combinatorial; therefore there is no need to bother about whether it is all about sentences or propositions. It should be noted, however, that Maimon maintained that what is true is primarily the sentence and not the thought. See *Versuch* 148.

10. Especially noteworthy in the discussion of Quine's conception of reduction are Steiner and Tharp. Steiner (1971) has pointed out that Quine's criteria for reduction do not exhaust the epistemic force found in classical reductions. The parsimony achieved in terms of ontological commitment is not the only purpose of reductions. Steiner proposes a condition on reduction to remedy this situation. In his view a good reduction is not a mere transference of a proof that retains its status as proof; it must also ascertain the systematic position of that proof, namely, its relations within its family. This is secured by transferring families of proofs that retain their family structure. Tharp (1971) also criticized Quine for contenting himself with a one-to-one correspondence from one domain to the other, preserving the language; confining ourselves to Quine's criterion, one can eliminate the mind-body problem in favor of a Pythagorean position. Tharp also notes Quine's overlooking that our theories are usually incomplete.

11. Indeed, it is by embedding in set theory that Paris (1977) proved a number of theoretical theorems that demonstrably cannot be derived from the Peano axioms.

12. As Maimon explains: If the mathematicians have proved, by way of example, familiar propositions about an angle or about cosine in general (whatever its size), then they also employ these propositions in the case of instances where these objects are of no magnitude at all (*Versuch* 353). See Buzaglo, *The Logic of Conceptual Change* (Cambridge, England: Cambridge University Press, forthcoming), wherein I study this process.

13. I prefer this example to that of the definition of the exponential function, based on the characteristic that the product of two powers (of the same base) is the

power whose exponent is the sum of the two respective powers. The advantage of this example is that the sinus function seemingly appeals to the intuition of a plane. The proof that these functions are determined uniquely by this characteristic is not difficult; for one can define a new function, $T(x) = F(x) + if(x)$, and prove on the basis of the defining characteristics of $G(x)$ and $F(x)$, which are specified in the main text, that $T(x)$ satisfies the exponential equation, whence the way to uniquely defining it over all fractions and through continuity over the irrationals is straightforward.

14. An interesting analysis of this geometry, in the context of Fichte's thought, can be found in the work of Gaston Bachelard (1949).

15. I do not wish to claim that this was indeed Klein's view on intuition in mathematics.

16. Cantor's research is hopelessly entangled with ontological commitments that Maimon could never take as they stand. I am appealing here to Cantor only in order to clarify an important aspect of the Maimonic notion of reduction, namely, the impossibility of exhaustively reducing intuition.

17. Quine in *Word and Object*, sec. 53, proposed a proximate position.

18. In fact, mathematicians see the Banach-Tarski paradox as symptomatic of the fault in the conception of the continuum. On the basis of this theory, we can prove that it is possible, by means of a complex algebraic-set-theory procedure, to divide a ball into four distinct balls of identical volume to that of the original ball. Many would hold that such a great deviation from our intuitions about volume indicates a basic problem in our set theory. This phenomenon leads us to conclusions as to the connection between the intuition about a continuum and the way in which we conceptualize it. Mathematical inquiry, by the very fact that it regards itself as having failed to capture certain intuitions, expresses its intention to conceptualize the intuition of a continuum. Just as, in the natural sciences, we modify theories in order to get closer to the essence of the subject of our inquiry, or disavow the validity of our feelings in preference of a coherent world, so it is concerning the study of an a priori object like the continuum.

Chapter Four

1. Maimon not only envisioned this possibility but also perceived the danger it poses to the Kantian position. For example, Cassirer, who speaks of mythic causality probably sees in these moves the fecundity of the Kantian thought.

2. Similar moves reemerged following the twentieth-century developments in logic and in the philosophy of logic. Wittgensteinian positions that logic is the condition for any thought, and Quine's later position about the implausibility of thinking beings that do not abide by the laws of logic, are related to this matter.

According to Quine's approach, a translation that does not conform to the rules of logic is not to be considered as such. Maimon's position on this point is more cautious. In Quinian terms, he does not argue that the very definition of translation requires the preservation of the laws of logic; it is sufficient for him that anyone who argues to this effect must explain what precisely constitutes a translation.

3. This principle has been preserved even in the theory of general relativity, notwithstanding the profound change the latter made in our conception of geometry. The new understanding of the straight line as a geodesic curve has preserved its metric properties.

4. If we adopt Bergman's interpretation of Maimon's critique of Kant, we may reach the same conclusion on the basis of Maimon's contention that by posing the question "How is experience possible?" Kant cannot possibly arrive at the transcendental deduction, because he has not substantiated the claim that objective experience is indeed accessible to us.

5. This distinction was brought into focus as a result of the work done by Kripke and Putnam.

6. Hume exerted a tremendous influence on the contemporaries of Maimon and Kant. For an instructive survey of this influence see Beiser (1987).

Chapter Five

1. Weinberg's comment that the world is a mathematical structure is completely in accord with the ideas mentioned here, providing this mathematical structure does not indicate intuitive geometrical terms.

2. Only resembles, since pure cognition includes more than logic.

3. I must admit that even the role of the context of discovery is not clear to me. Why should a picture in the mind play a role in making the discovery, while the table next to me has no such role?

4. Apparently, Frege relates to this question when he explains that "were it not possible for man to think and to make as the object of his thinking, things that are not carried within himself, then he would have an internal world but not a surrounding world" ("Thoughts" 27). A necessary condition for relating to the world that crosses one's inner world is that it becomes possible "to make as the object of one's thinking" the things that are outside him or her. But this comment is at most an answer to the quid facti question—humans think and thereby they relate to those things that do not depend on them. But the question that interests Maimon is the quid juris question—namely, how can we make intelligible the ability of human beings, who are flesh and blood, to perceive anything that is outside themselves.

5. Another direction that should be mentioned in this context is to hold that

intuition is conceptual, without saying it is potentially intellectual, and thus to maintain the epistemic importance of intuition in a realistic world, without becoming Platonistic (which sees material objects as manifestations of pure entities). This direction can be read in McDowell's (1998) version on the relation between intuitions and concepts.

6. This move is better than Frege's assumption that although we justify geometric axioms by invoking intuitions, other creatures may justify these propositions on different intuitions.

7. This argument is different from the one discussed in the literature, which questions the transition from "*A* is a condition for *B*" to "*A* is not experiential."

8. This doubt supports Hintikka's position, which understands the concept of intuition as a representation that does not necessarily refer to perception in time and place.

9. In view of this, it is not true that, as suggested by Allison (1983, 60), the reliance on the concept of infinite understanding makes Maimon's system theocentric. The notion of the infinite understanding, as suggested by Maimon, is purely an idea.

Chapter Six

1. For a brief survey of the literature on this issue, see Bergman (1967, 238–47).

2. Bergman (1967, 29) uses this idea of Maimon to explain the notion of intentionality in Brentano's philosophy in Ges. Werke, vol. 3, 1-246.

3. On this see *Philosophisches Woertebuch,* 176.

4. This is not because they are subjective. Even if the notion of line was a simple one, it would not be a representation.

5. According to Bergman's interpretation, Maimon did not question the value of the transcendental deduction (1967, 82), but only criticized Kant for assuming that we possess objective experience. This reading implies that the problem that Maimon saw in Kantian system is the *quid fact;* to this I cannot agree.

6. See the beginning of the following chapter.

7. One possible source of this view may be found in the kabbala (see the epilogue).

8. An elaboration of this idea is found in Atlas (1964, 63–108).

Chapter Seven

1. The relation between a species to its genus was ignored by modern logicians. I am developing a logic of this relation.

2. The relation of determinability is central to Fichte and Hegel. Some expressions of Fichte's and Hegel's were taken verbatim from Maimon (see Bergman 1967, 164, and his chapter on Hegel and Maimon).

3. Maimon, as I read him, does not agree with Kant (A-571, B-599) on this issue.

4. There is another problem with the very thought that cause and effect are intelligible relations that one can, according to Kant, deduce from logic: "Kant deduces the concept of cause from the logical form of hypothetical propositions. However, one may ask how it is that logic actually came to have this peculiar form, namely, that if one posits one thing, one must necessarily posit another . . . we do not encounter the hypothetical form anywhere, for we ascribe the properties of essence and the subject's predicates in an absolute manner. And even if this decisive proposition can be expressed hypothetically, it is the expression of the proposition, and not the form itself, which is hypothetical" (*Versuch* 183). Maimon is referring here to the possibility of interpreting a categorical proposition of the form "All *A* is *B*" as the hypothetical proposition: "If something is *A* then it is *B*." This analysis, which is close in spirit to the Fregean conception of categorical prepositions, is not sufficient to elicit the kind of conditionals associated with causality. Indeed, in mathematics, the general propositions have nothing to do with causality. We can say "If a line is straight, then it is the shortest distance between two points," but in making this claim we are in actual fact referring to the claim that every straight line is the shortest distance between two points (*Versuch* 184).

Predicate calculus, which came later, regards categorical propositions as hypothetical propositions, but this cannot provide us with a reply to Kant, for Kant would not have interpreted the general connection as a material implication. Naturally, Maimon does not formulate this in such terms, but his words imply that a general proposition lacks the inner connection necessary for Kant's deduction of causality. In the absence of another explanation, Maimon entertains the possibility that the kind of hypothetical proposition on which Kant bases his reasoning originates in the permeation of causality into the field of logic and therefore it has to be removed from logic instead of being deduced from it.

5. Quine's idea of indeterminacy of translation and Goodman's new riddle of induction are two examples.

6. Here lies an important difference between Maimon and the neo-Kantians not recognized by Atlas (1964, 330).

Epilogue

1. The simplest example that can be given to illustrate this idea is the way we define complex numbers in algebra and locate them in Gauss's plane.

2. *Matter* here is not in the Aristotelian sense.

3. Maimon would not have argued that God does not exist. Nor would he have argued that God does exist. Concerning this issue, he accepts the doctrine of the negative attributes, which removes the attribute of existence from God. In the entire course of his writing, Maimon did not withdraw from this position. The existence of God as an ideal is Maimon's own interpretation of Maimonides's theory of the divine negative attributes. This interpretation would not have been accepted by Maimonides himself, for whom divine existence is unique because God is a necessary being.

4. Consult Idel's paper "From Jewish Isoterics to European Philosophy" (chap. 8 of Idel 1988).

5. Recently it became apparent that Leibniz had studied kabbala seriously and was influenced by it. See Coudert 1995.

6. We are indeed acquainted with Kant's mockery of Swendenborg's "Dreams of a Spirit-Seer" (1766); but precisely this mockery, just like his making fun of the kabbala (chap. 3 of this work is called "anti-Kabbala") attests to his tremendous attraction to this realm of thought. In fact, before Kant conveyed his mordant criticism on Swendenborg's work, he wrote him a letter in which he expressed his high esteem.

7. I am grateful to my friend Mr. Yoram Navon, who helped me bring into focus this aspect of Maimon's activity.

Appendix

1. I have expanded on this in my book *The Logic of Concept Expansion* (Cambridge, England: Cambridge University Press, forthcoming).

2. For Maimon, an expression like $\sqrt{2}$ is an idea of reason, since we regard a convergence as an object, but we know that bona fide numbers are either fractions or integers. However, although the expression $\sqrt{(-1)}$ is also an absurd expression, Maimon didn't regard it as an idea of reason, since the illusion of convergence to this number (as it is for irrational numbers) doesn't occur here. The point is that the system of (all) natural numbers can't exist for a finite creature, and in fact contains a contradiction. This is one major difference between Maimon and Dedekind, who saw no problem in accepting an infinite set that contains all possible objects of thought.

3. It would seem that he assumed that whenever there is a concept, to which no intuition corresponds, there must be some relation to infinity; from the development of mathematics we know that this conception is not accurate. Klein's bottle is a clear concept without a corresponding intuition, and it involves no issue of

infinity. My purpose here is not an exposition of Maimon's answer to Kant's challenge. Such an exposition would take us to Maimon's doctrine of the fiction and would detract us from the *quid juris* question.

4. One ought not conclude from this that ideas of reason do not have their corresponding intuitions. Maimon's solution to the *quid juris* question is that intuitions, as such, are ideas of reason. This was expanded to emprical intuitions and entails that the difference between Caesar and 7 is not the difference between an object with inner properties and an object that belongs to a structure; rather, it is the difference between two objects that belongs to different structures.

5. It would be interesting to read the quotation from the *Versuch* (78) as an intuitionist gloss on the universal quantifier.

6. For Maimon, since the sequence of mutual determinations is made by the subject, it is impossible to apprehend all of them. The structuralist aspect of his view only makes things more difficult: We don't have 2 if we don't know its relations with 3, 4, etc. So we never actually apprehend any number. Maimon, however, accepted the use of ideas of reason despite their inherent contradictions. Here is a tricky way of dealing with this problem for the set of the integers Z: Let us take different finite fields, of the type Z_p; e.g., Z_7, Z_{11}, Z_{13}, etc. Each of these structures is finite, so each is a bona fide object. The limit of this process, so to speak, is Z, the set of integers.

Bibliography

Allison, H. E. *Kant's Transcendental Idealism.* New Haven, Conn.: Yale University Press, 1983.

Angelelli, I., ed. *Begriffsschrift und andere Aufsätze.* Hildesheim, 1964.

Atlas, S. *From Critical to Speculative Idealism.* The Hague: Martinus Nijhoff, 1964.

Bachelard, G. *Le Rationalisme Applique.* Paris: Quadrige, 1949.

Beiser, F. C. *The Fate of Reason.* Cambridge, Mass.: Harvard University Press, 1987.

Benacerraf, P. "What Numbers Could Not Be." 1965. In *Philosophy of Mathematics,* ed. P. Benacerraf and H. Putnam. Cambridge, England: Cambridge University Press, 1983.

Bergman, S. H. *The Philosophy of Solomon Maimon.* Jerusalem: Magnes, 1967.

Beth, E. W. "Über Lockes 'allgemeines Dreieck.'" *Kant-Studien* 48 (1956–57): 361–80.

Bransen, J. *The Antinomy of Thought: Maimonian Skepticism and the Relation between Thought and Objects.* Dordrecht: Kluwer, 1991.

Buchdahl, G. *Kant and the Dynamics of Reason.* Oxford and Cambridge: Blackwell, 1992.

Buzaglo, M. *The Logic of Concept Expansion.* Cambridge, England: Cambridge University Press, forthcoming.

———. "Quasi-Logicism" (Hebrew). *Iyyun* 46 (April): 115–37.

Cassirer, E. *Das erkenntnis problem.* Berlin: Cassirer, 1920.

———. *Substance and Function.* Chicago: Open Court Publishing, 1923.

Coudert, J. A. *Leibniz and Kabbalah.* The Hague: Kluwer, 1995.

Dedekind, R. "Was sind und was sollen die Zahlen" (The nature and meaning of numbers). 1888. In *Essays on the Theory of Numbers,* ed. W. W. Bennan, 31-115. New York: Dover, 1963.

Dummett, M. *Frege and Other Philosophers.* Oxford: Clarendon Press, 1991.

———. *Frege: Philosophy of Mathematics.* London: Duckworth, 1991.

Engstler, A. *Untersuchungen zum idealismus Salomon Maimons.* Fromman-Holzboog, 1990.

Fichte, J. G., *Science of Knowledge.* Trans. Peter Lauchlan Heath. New York: Appleton-Century-Crofts, 1970.

Frege, G. *Collected Papers.* Oxford: Basil Blackwell, 1984.

———. *The Foundations of Arithmetic.* Trans. J. L. Austin. Oxford: Blackwell, 1950.

———. *Grundesetze der Atritmetik.* In *Translations from the Philosophical Writings of G. Frege,* ed. P. Geach and M. Black. Oxford: Basil Blackwell, 1966.

———. "Thoughts." 1918. In *Logical Investigation.* Trans. P. T. Geach and R. H. Stoothaff. Oxford: Basil-Blackwell, 1977.

———. *Philosophical and Mathematical Correspondence.* Chicago: University of Chicago Press, 1980.

———. *Posthumous Writings.* Chicago: University of Chicago Press, 1979.

Friedman, M. 1985 "Kant's Theory of Geometry." Reprinted in *Kant's Philosophy of Arithmetic,* ed. Carl Posy, 177–221. The Hague: Kluwer, 1992.

Gueroult, M. *La Philosophie transcendentale de Solomon Maimon.* Paris: F. Alcan, 1929.

Harper, W. "Kant on Space, Empirical Realism, and the Foundations of Geometry." In *Kant's Philosophy of Arithmetic,* ed. Carl Posy, 257–93. The Hague: Kluwer, 1992.

Hintikka, J. "Are Logical Truths Analytic?" *Philosophical Review* 74 (1965): 178–203.

———. "On Kant's Notion of Intuition (Anschauung)." In *Kant's First Critique,* ed. Terence Penelhum and J. J. MacIntosh. Belmont, Calif.: Wadsworth, 1969.

———. "Kantian Intuitions." *Inquiry* 15 (1972): 341–45.

———. "Kant on the Mathematical Method." *Monist* 51 (1967): 352–75.

Idel, M. *Kabbala: New Perspectives.* New Haven, Conn.: Yale University Press, 1988.

Kant, I. *Critique of Practical Reason.* Trans. Lewis White Beck. New York: Library of Liberal Arts, 1956.

———. *Critique of Pure Reason.* Trans. N. Kemp Smith. Houndmills, U.K.: Macmillan, 1933.

———. *Dreams of a Spirit—Seer, 1766.* In *Immanuel Kant, Theoretical Philosophy, 1755–1770.* Ed. and trans. D. Walford. Cambridge, England: Cambridge University Press, 1992.

———. *Kant's Introduction to Logic.* London: Longmans Green, 1885.

———. *Philosophical Correspondence, 1759–1799.* Ed. and trans. A. Zweig. Chicago: University of Chicago Press, 1967.

Kuntze, F. *Die philosophie Solomon Maimons.* Heidelberg: Inter, 1912.

Leibniz, G. *Die philosophische Schriften von G. W Leibniz.* I–VII. ed. C. I. Gerhardt. Hildesheim: Halle, 1971.

———. *Mathematische Schriften.* I–VII. Ed. C. I. Gerhardt. Hildesheim: Halle, 1971.

Maimon, S. *Gesammelte Werke.* I–VII. Hildesheim: Georg Olms Verlagsbuchhandlung, 1965.

———. *Givat Hamoreh* (Hebrew). Magness, Jerusalem: Hebrew University, 1967.

———. "Heshek Shelomo." National Library Archives, Jerusalem.

Maimonides, Moses. *Guide of the Perplexed.* Trans. Shlomo Pines. Chicago: University of Chicago Press, 1963.

McDowell J. *Mind and World.* Cambridge, Mass.: Harvard University Press, 1998.

Parsons, C. "Arithmetic and the Categories." 1984. In *Kant's Philosophy of Arithmetic,* ed. Carl Posy, 135–58. The Hague: Kluwer, 1992.

———. "Kant's Philosophy of Arithmetic." 1969. In *Kant's Philosophy of Arithmetic,* ed. Carl Posy, 43-68. The Hague: Kluwer, 1992.

———. "Mathematical Intuition." *Proceedings of the Aristotelian Society* (1979–80): 145–68.

———. "The Structuralist View of Mathematical Objects." In *Mathematical Objects and Mathematical Knowledge,* ed. M. D. Resnik. Aldershot, U.K.: Dartmouth, 1995.

Posy, C. J. "Kant's Mathematical Realism." In *Kant's Philosophy of Mathematics,* ed. C. J. Posy. Dordrecht: Kluwer, 1992.

Putnam, H. *Philosophical Papers.* Vols. 1–3. Cambridge, England: Cambridge University Press, 1975–83.

Quine, W. V. *From a Logical Point of View: Logico-Philosophical Essays.* Cambridge, Mass.: Harvard University Press, 1961.

———. "Ontological Reduction and the World of Numbers." Cambridge, Mass.: MIT Press, 1964.

———. *Word and Object.* Cambridge, Mass.: MIT Press, 1960.

Reichenbach, H. *The Philosophy of Space and Time.* New York: Dover, 1958.

Resnik, M. *Mathematics as a Science of Patterns.* Oxford: Oxford University Press, 1997.

Rotenstreich, N. "On the Position of Maimon's Philosophy." *Review of Metaphysics,* 21 (1968): 534–45.

Russell, B. *Mysticism and Logic.* London: Penguin Books, 1918.

Sluga, H. "Frege's Alleged Realism." *Inquiry* 20 (1977): 227–42.

———. *Gottlob Frege.* London: Routledge and Kegan Paul, 1980.

Steiner, M. *Mathematical Knowledge.* Ithaca: Cornell University Press, 1975.

———. "Quine and Mathematical Reduction." In *Essays on the Philosophy of W. V. Quine,* ed. R. W. Shahan and C. Swoyer. Norman: University of Oklahoma Press, 1979.

Strawson, P. F. *The Bounds of Sense.* Oxford: Clarendon Press, 1970.

———. *Individuals.* London: Methuen, 1959.

Tharp, L. H. "Ontological Reduction." *Journal of Philosophy* 68 (1971): 151–64.

Thompson M. "Singular Terms and Intuitions in Kant's Epistemology." In *Kant's Philosophy of Mathematics,* ed. C. J. Posy. Dordrecht: Kluwer, 1992.

Tishby, I. *The Wisdom of the Zohar.* Trans. D. Goldstein. New York: Oxford University Press.

Vaihinger, H. *The Philosophy of "As If."* New York: Harcourt, Brace, 1925.

Wolff, C. *Preliminary Discourse on Philosophy in General.* New York: Bobbs-Merrill Co., 1963.

Index